Lexical Processes in Scientific Discourse Popularisation

english corpus linguistics

Thomas Kohnen · Joybrato Mukherjee (eds.)

Volume 5

PETER LANG
Frankfurt am Main · Berlin · Bern · Bruxelles · New York · Oxford · Wien

Christiane Brand

Lexical Processes in Scientific Discourse Popularisation

A corpus-linguistic study of the SARS coverage

PETER LANG
Internationaler Verlag der Wissenschaften

Bibliographic Information published by the Deutsche Nationalbibliothek
The Deutsche Nationalbibliothek lists this publication in the Deutsche Nationalbibliografie; detailed bibliographic data is available in the internet at <http://www.d-nb.de>.

Zugl.: Gießen, Univ., Diss., 2006

D 26
ISSN 1610-868X
ISBN 978-3-631-56012-9

www.peterlang.de

Acknowledgements

This book would not have been possible without the support and encouragement that I have received from many people. First and foremost, I owe a great debt to my supervisor Professor Joybrato Mukherjee. He has not only dedicated much time and effort in reading and commenting on various drafts, but he has also inspired and guided me all along the way. Personally, I have benefited greatly from his expertise and academic insights. It was Professor Mukherjee who provided me with direction and put me firmly on the linguistic track. I am most thankful for this. I would also like to express my gratitude to Professor Magnus Huber who has always supported me with good advice, warm words and hot tea. Professor Matthias Hutz and Kathryn Khairi-Taraki already believed in me when I was still a student and encouraged me to start this project in the first place. A very special thanks goes to Rosemary Bock who proofread the manuscript in the most competent way and has made many valuable suggestions about popularisation. I am indebted to all my colleagues in the linguistic department, especially to Marco Schilk, Sandra Götz and Carmen Aguilera Carnerero who always gave generously of their time in order to discuss corpus-linguistic issues and to share essential daily cups of coffee with me. Most importantly, I owe a great debt to my family and friends. Rebecca Fabian's skilled feedback was vital for many parts of this book. Of course, I would like to express my deepest gratitude to my parents. My mom has supported me in every possible way and it was my dad's vision that made me believe in what I was doing. This work is dedicated to him.

Table of Contents

1 Introduction

1.1 Motivation and objective of the present study

In 2003, there was an unprecedented health crisis caused by an unknown virus that had emerged in Asia. The new disease soon became known as SARS (Severe Acute Respiratory Syndrome). It was initially discussed in medical circles and on a scientific basis only, for example, via articles in journals, such as the British Medical Journal. After the first signs of a worldwide epidemic, the subject was taken up by the general newspress and was henceforth covered on a large scale. At that stage, the subject-specific discourse was simultaneously realised on two distinct levels: the scientific and the popular genre. Set within the framework of the popularisation of science, the present study focuses on similarities and differences of both genres on the lexical level. I will investigate how a scientific topic is presented in the written media of mass communication that is outside the realm of science itself. I seek to answer the questions of what is communicated in what way and what modes are involved on the lexical level. Can lexical structures give evidence of interrelation between the two genres or are the two realisations independent varieties?

1.1.1 Scientific discourse popularisation

In order to meet the demands of today's informed society, the news media has taken it on itself to provide the general public with information that is technical in nature and which was previously restricted to the scientific discourse community. As a result, the scientific discourse has now a popular counterpart. Yet, the study of popularisation is not only concerned with the question how science information is communicated to a lay audience, but is furthermore interested in implications for a more general communication process. It is a fairly recent development in Language for Specific Purposes (LSP), but, as Weigold (2001) claims, it is all set to have a bright future:

> This healthy and vibrant area of scholarship is likely to become even more central in the discipline of mass communication. The special challenges presented by communicating the complex and important issues of science will only grow in importance in this new century. (Weigold 2001: 187)

Although science popularisation was originally studied for pedagogical purposes, the focus today is on science news, that is on how science gets transformed from research results into news stories. In this study, I intend to question the boundaries of scientific discourse and genres of popularisation. Earlier studies either regard the process of popularisation as strictly linear, with the popular discourse being a mere simplification of its scientific source, or claim that there is a vast gulf between the two forms and that they are stories in their own right[1]. Recently, researchers have shown interest in a more multidimensional approach, examining the way in which both discourses interact.

The SARS crisis coverage in 2003 developed between scientists and the news on a global scale. SARS is an infectious disease that is facilitated by the mobility of today's society and as global travel continues to grow, the worldwide spread of such viruses and bacteria becomes more and more rapid. As a consequence, not only Asia, where the virus emerged in February 2003, but also the rest of the world was affected by the disease. As cases started to appear in Europe, Canada and Australia, the topic became a key issue in public life. The SARS discourse was then realised at two different - yet parallel - levels: medical scientists discussed the disease within the confines of their special field, whereas the general public developed their own approach to the topic in newspapers and magazines. This dual nature can be best described by looking at the different levels of the specific discourse. Table (1) illustrates the differences in (written) forms of discourse which are nevertheless connected in that they pass on information about one and the same topic:

Table 1: *The two levels of the SARS discourse*

Topic	Level	Audience	Objective	Medium
SARS	internal	restricted; medical professionals with special education and knowledge	presentation and discussion of relevant facts	research article or bulletin in Medical Journals
	external	general; public at large without field specific education and know-ledge	popular broadcast of the events, infotainment	news article in papers and magazines

Communication at the internal level takes place between experts exclusively within the field. Both the authors and the recipients have received a particular education and gained qualifications in this field so that they are of course familiar with the highly codified medical language and terminology. The

[1] For a critical overview of present popularisation studies see Myers (2003).

scientific corpus compiled for this study consists of Medical Journal articles written by medical professionals whose objective is to forward clinical results and practical realisations about the disease to their colleagues.

External communication is intended for laypeople. It manifests a discourse about scientific topics outside the respective professional field, for example, via the news press. Therefore, the popular corpus compiled for this study comprises newspaper and magazine articles. It has to be stressed at this point, that journalists take a mediating position in the public SARS discourse, but they belong, in the strict sense, to the medical laity. The purpose of the articles included in the popular subcorpus is to create awareness and to inform about the topic in the genre specific infotaining style that presents scientific facts in an entertaining and appealing way. I argue that this duality must be reflected in language use and this study sets out to explore certain lexical characteristics by comparing the two subcorpora.

1.1.2 A corpus linguistic approach

The use of electronic corpora in LSP research has expanded within the last 10 years. Former studies drew evidence from paper-based collections rather than from electronic databases. Biber (1988) and Crystal (1991) call for a corpus approach to register analysis rather than the individualistic and intuitive way of interpreting single data. In Figure (1) Aarts (2002) refers to the issue of intuition and other linguistic data, including corpora. As illustrated in Figure (1), Aarts shows that there are different kinds of linguistic data and that for a long time linguists had to rely on their own intuition when they made assumptions about language rules and patterns. The introduction of corpus linguistics and computer-assisted corpus analysis, made it easier to provide a source of evidence by drawing on authentic and non-intuitive data. According to Hunston (2002: 1) corpora allow linguists not only "to count categories in traditional approaches to language but also to observe categories and phenomena that have not been noticed before". Corpus linguistics and Language for Specific Purposes (LSP) mainly meet in the restricted context of identifying and studying individual field-specific or genre-specific patterns. Up to now, the lexicon of special languages has attracted the attention of scholars interested in fields, such as phraseology, lexicography and terminology research (cf. Cohen 1986, Kjaer 1990). Most recently, the contrastive analysis of different special languages or genres has also become of interest. Biber et al. (1998, 1999), for example, investigate the temporal use across conversational, fictional, news and academic registers by comparing frequencies.

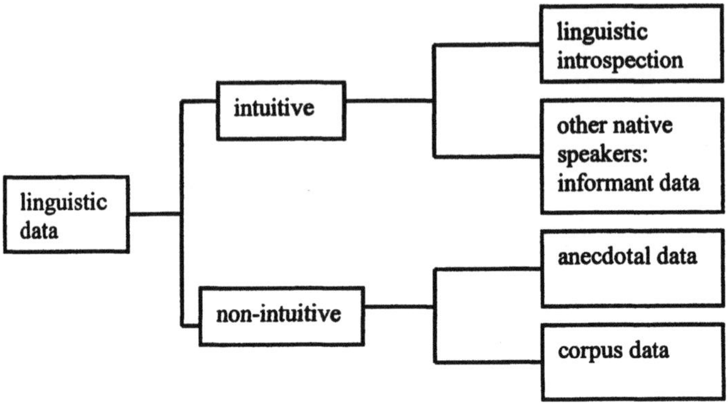

Figure 1: *Intuition and corpus data (Aarts 2002: 4)*

In this study, I intend to shed light on a rather new perspective by applying corpus-linguistic methods to the study of science popularisation, i.e. the research questions demand a fairly specialised corpus. I have thus decided to use a corpus that I have especially compiled for this study rather than choosing a particular subsection of an already available large corpus like the BNC. The case of SARS provides an exceptional opportunity as the corpus could be extracted on-time parallel to the actual events. It is thus comparable in topic, time frame and medium. The resulting two corpora provide a solid basis for the computer-assisted exploration of the data.

1.2 Preview

This study offers a comparative investigation of lexical patterns used in the scientific context of written medical communication and related public news coverage. In order to create a coherent picture of both levels of communication, Chapter 2 discusses the nature and development of Language for Specific Purposes (LSP). Special attention will be paid to the lexical level, i.e. to terminology and to the concept of horizontal and vertical division. In the following section, the discussion of popularisation takes up the idea of verticality and links it to communication barriers. How these communication barriers are bridged is at the heart of the theory of popularisation outlined in Chapter 2 in which the origin, development and main objectives of science popularisation are outlined and current approaches and strategies are discussed in more detail. A look at science journalism then provides some more insights into news discourse. In order to analyse the different lexical realisations of specific concepts in each genre (popular and scientific), a corpus of 500,000 words has been compiled, consisting of medical journal and general newspaper articles.

Chapter 3 discusses some current issues in corpus theory and their relevance to the present study, such as the corpus-based vs. corpus-driven approach, and gives insights into the compilation process. The corpus design is explained and the methodology outlined. Afterwards, a pilot study explores the corpus and puts the analytical procedure to a first test.

The quantitative analysis in Chapter 4 aims at illustrating the interrelation between the popular and the scientific texts by comparing their lexical content on a frequency basis.

This is complemented by the discussion of selected concordance lines in each genre in Chapter 5. The qualitative analysis sets out to examine not only the context of the respective lexical items, but furthermore sets collocations and clusters across the corpora in relation to each other.

Chapter 6 looks in more detail at their use and thus the relation between form and function. Variables of word choice are identified and a model of lexical realisations in scientific discourse popularisation is lined out.

Finally, implications of the findings, as well as suggested paths for future research, will be discussed in Chapter 7.

2 Theoretical background

2.1 LSP: Language for Specific Purposes

At the beginning of the last century, significant progress in the domains of science and technology opened up the way to a new view on particular language means that were required to deal with the rise of *Fachsprache* or *Language for Specific Purposes*. Although the tradition of *Fachsprachenforschung* was originally established in central Europe, the growing interest in the study of special languages in the English speaking world soon led to the coinage of the term *Language for Specific Purposes* or LSP in short. It had been chosen as an equivalent for the already existing German name and concept of *Fachsprache*, bringing the incipient process of borrowing the Germanic word to a halt. As a consequence, the expression *Fachlichkeit* has no consistent equivalent in English: *Fachmann* is referred to as *expert* or *specialist*, *Fachsprachenforschung* as *LSP research* and *Fachlichkeitsgrad* can be best described as *level* or *degree of specialisation*. Yet, the basic concept of *Fach* is labelled *domain*, *field* or *discipline*. I will use the terms *LSP* and *special language* in referring synonymously to one and the same concept of *Fachsprache*.

Linguists agree on defining *Language for Specific Purposes* as a complex domain of language use determined by the specification of the scientific or technical context. It is applied by educated, and thus competent, speakers or writers in order to communicate with other representatives of the same, or of neighbouring disciplines (cf. Beier 1980; Hoffman 1990; Sager et al. 1980, Roelcke 2005). Although LSP research as an established linguistic field is still a relatively young discipline, preceding interests in the manifestation of language in professional and academic settings have led to this development.

In the following sections, I will first provide an overview of the most significant stages in LSP development and look at the main characteristics and the various domains of special languages in use.

2.1.1 Development and state of the art in LSP research

The concern about special language and, first and foremost, about its specific terminology has grown out of time and is embedded in the historical development of science itself. The structure and vocabulary of academic disciplines is clearly shaped by their classical Greek and Latin roots. The earliest systematic collection of medical terms, for example, dates back to Hippocrates, who composed the first written documentation of human pathology, illness and medication.[2] For centuries, reflection about the special language was restricted to scientists and professionals of the respective domains. They regarded their language as a key tool which had to be developed "so that it could be properly used for those tasks in which it was necessary to avoid the ambiguity and lack of precision of everyday speech" (Laurén 1994: 87).

It was not until the beginning of the last century that aspects of special languages were discussed outside the individual field. From the 1930s onwards, two linguistic approaches concerned with LSP emerged: terminology research on the one hand and *Wirtschaftslinguistik* on the other. The history of terminology research is closely linked to the Austrian scholar Eugen Wüster. He initiated systematic terminology research with his work on language standardisation in technology (cf. Wüster 1931) and established terminology and terminography as an independent discipline. Wüster and his later colleagues at the University of Vienna framed a theory that considers terminological concepts as central elements of special language. They furthermore stressed the importance of transparent definitions for those concepts which can be transferred from one language to the other without losing the explicit reference. The instruments created for terminology research and concept analysis served as central tools for the following (descriptive) studies of special languages.

One of the oldest branches of special language research, *Wirtschaftslinguistik*, is regarded as the predecessor to modern business communication and was established to feed the demand for appropriate education expressed by the then flourishing schools of commerce (*Handelsschulen*) in Europe. Ewald Messing was one of the pioneers in this field. He initiated the linguistic discussion of the topic by publishing an anthology titled *Wirtschaftslinguistik* in 1932. The work was not only dedicated to teaching business terminology, but, according to Drodz & Seibicke (1973), it was also shedding new light on linguistics by considering factors, such as cultural awareness and communicative study of culture (*Völkerverständigung*). Another special interest in this perspective on special language was to investigate the relation between economic progress and the development of language itself. Those investigating *Wirtschaftslinguistik* pointed out that this overall approach to language and communication could be seen as one of the precursors of applied linguistics in Europe (cf. Drodz & Seibicke 1973, Gerzymisch-Arbogast 1987).

At the same time, the structural and functional language theories introduced by members of the Prague school also influenced the view on special languages considerably. The theories claim that linguistic variation is a direct result of the need to fulfil particular expressive and / or stylistic functions. This intellectualisation of language thus requires an individual and functional use of lexical and syntactic devices. Since the functional view of language forms a

[2] *The Hippocratic Collection*; 5th\4th century B.C.

major part of the framework supporting this study, I will not discuss those concepts in this section but take a closer look at the theories in part 2.3, which deals with the Prague school and related concepts.

The first general Anglo-Saxon linguistic approach to LSP was concerned with the *Linguistic Aspects of Science* as the title of the first publication on this subject matter by Leonard Bloomfield (1939) indicates. Bloomfield did not look at specific vocabulary in isolation, but presented his overall impression on the extraordinary properties of language in scientific settings. His initial observations addressed the art of mastering this special form in spoken discourse:

> The use of language in science is specialized and particular. In a brief speech the scientist manages to say things which in ordinary language would require a vast amount of talk. His hearers respond with great accuracy and uniformity. The range and exactitude of scientific prediction exceed any cleverness of everyday life.
> (Bloomfield 1939: 1)

Bloomfield describes the *peculiarities* of scientific language (as he calls them) and puts the reflection on LSP in line with dialectology, appealing to his colleagues that "[i]n all this development we have not left the domain of language. We reach its outskirts." (Bloomfield 1939: 44). However, this call for attention did not raise notable interest in the linguistic community and studying special languages was rather left with the respective professionals and terminologists.

Most of the approaches outlined so far came to an unfortunate standstill in the course of World War II. It was not until the 1970s that a school of thought emerged in Leipzig (former German Democratic Republic), which finally established the study of LSP (*Fachsprachenforschung*) as a distinct linguistic discipline. In their study of special language, the scholars in Leipzig were at first notably influenced by the functional and stylistic view of the Prague school. It was Rosemarie Gläser who carried out the first comprehensive study of stylistic components in LSP. Her work *Fachstile des Englischen* (1979) is still considered as fundamental. She discusses major stylistic devices at the lexical, phraseological, syntactical and textual level and sets the linguistic means into a broader context by relating them to intention, form and effect of the actual utterance:

> Fachstil ist die für die Gestaltung eines Fachtextes charakteristische Auswahl und Anordnung sprachlicher Mittel, die in einem Gesamtzusammenhang von Absicht, Inhalt, Form und Wirkung der Aussage fungieren. (Gläser 1979: 26)

Later, Gläser extended her studies of text and genre analysis and left the narrow framework of functional stylistics by integrating textlinguistic as well as pragmatic approaches into her work on text typology (cf. Gläser 1979, 1990, 1992, 1995). This transition was further encouraged by fellow scholars in Leipzig. Lothar Hoffmann proposed not only the study of lexical patterns as well as grammatical devices, but took also the leading role in establishing the communicative approach in LSP research. His book *Kommunikationsmittel Fachsprache - Eine Einführung* (Hoffmann 1985) served as the starting point for the general discussion of LSP in a broader linguistic setting. In the 1980s, this progress led to what is today known as the *communicative turn* (cf. Helbig 1986). Methodologically, the focus shifted from small lexical or grammatical items to entire texts and other forms of discourse. De Beaugrande and Dressler (1981) define this progress adequately as a shift from one question to another, that is from: Which linguistic structures can we discover in our analysis? to: How do we create structures through operations of decision and choice and what are the implications of such operations for communicative interaction? Hence, substantial outcomes of the *Leipziger Fachsprachen Studien* are not only indispensable publications and numerous conferences and colloquia, but also concepts that are still valid today. These concepts offer applications and implications for present LSP research.

Apart from the Leipzig school, one also has to take into account the continuous efforts elsewhere made from the 1980s onwards that aimed at describing and characterising individual special languages or *technolects* (cf. Beier 1980, Fluck 1991). The profiles that were constructed are those of established special language varieties, such as the *Language of Law* or *Medical Language*, but also include techno- or minilects, such as the Language *of Fishing* or *AirSpeak* (the language used in air-traffic control). The purpose is, on the one hand, to investigate the structure of those varieties, on the other hand, also to draw practical conclusions for the professionals using this type of special language. Johnson (1993), for example, decided on analysing *PoliceSpeak* because linguistic counselling "would make police communications more concise, more predictable, more stable and less ambiguous" (Johnson 1993: 739). Linguistic counselling of course refers to special language didactics, which plays an important role especially in the English speaking world, where the notion of LSP was soon replaced by the significance of ESP (English for Specific Purposes), as the following paragraph will illustrate.

Apart from Europe, the study of LSP in the U.S.A and the UK was shaped by scientific discoveries and first and foremost by technological innovations. In Washington a school of thought was founded by Louis Trimble in the 1970s, focusing on prescriptive studies that had practical implementations for the occupational and technical domains. Louis Trimble, Mary Todd Trimble and Larry Selinker were the first who specialised in teaching EST (English for

Science and Technology) to advanced students and gained their expertise in this specific field out of "sheer need":

> We very quickly realized that before we could adequately teach students the English of Science and Technology we had to learn something about it. Our research began with an effort to determine the essential nature of scientific and technical English by finding its major characteristics and where it differed from other forms of written English. (Trimble 1985: 1)

They consequently pursued the idea of providing (international) students of English with a customised form of education that they would not receive in a standard EFL course. Trimble and Selinker concentrated on studying linguistic features of engineering and science texts with a dual aim: They intended to meet the occupational requirements of their students, but were at the same time concerned with the integration of the linguistic study of different registers.

The rise of English as the *lingua franca* in the 1980s and 1990s encouraged the worldwide development of ESP studies. English is nowadays, in the age of globalisation and the new media, the unchallenged *lingua franca* not only of science but of numerous academic and professional fields. It has apparently taken over the position that Latin once possessed various reasons. Chargaff gives a simple explanation: "[T]he English language is much easier to access, but at the same time also much easier to use badly" (Chargaff 1986: 109). Nevertheless, ESP today sees its role not only in teaching students how to use technical or scientific English in an acceptable way but also in handing over relevant skills, such as communicative competence and (inter-) cultural sensitivity.

So far, I have sketched out the major approaches and stages of LSP in the course of the last century, but what are the applications and challenges of current LSP research? There is of course the ongoing concern about special language didactics and ESP. Apart from the prominent aspect of intercultural competence mentioned above, the major discussions are about methodological issues. Computer aided learning and the use of corpora are undoubtedly on the high rise. The concept of *data-driven learning*, or *DDL* for short (cf. Johns 1991), has developed out of the learning-centred approach in foreign-language teaching. It employs the notion of inductive *discovery learning* (Aston 2001:19) which encourages learners to explore lexicogrammatical structures. Such activities are still the exception in today's special language didactics. Lynne Bowker and Jennifer Pearson were first to publish a textbook on working with specialised corpora in 2004. It aims not only at advanced students of linguistics but also at students of the respective professional fields and outlines the concepts, compilation and analysis of LSP corpora and encourages the students to explore a corpus in order to study, e.g. multi-word phrases or to track meaning out of the context.

A further key issue of LSP is the concern about cognitive aspects. Baumann (2001) even proclaims the ***cognitive turn*** in LSP, stating that the joint efforts of linguistics and cognitive science are opening up new horizons in the understanding of the complexity of special languages:

> The development of the present LSP research in the direction of cognitive science has caused a broadening and deepening of its subject-matter. Now it explores the different communicative-cognitive procedures of information transfer and processing comprising all possible levels of LSP communication. (Baumann 2001: 89)

Baumann mentions the connection between cognitive studies in LSP and the concepts of information transfer and processing. The growing interest in information transfer and knowledge engineering has led to a number of colloquia and publications on the topic of *Wissenstransfer* (cf. Wichter & Antos 2001; Hoffmann, Kalverkämper & Wiegand 1998). These recent additions to LSP research draw on the complex mental representation of specialised knowledge and aim at explaining the processes involved in information storage and transition. This is directly linked to the question of specific matters are passed on to serve either educational or informative purposes. From the 1990s onwards, this development has attracted more and more interest and, as will be outlined in detail in chapter 2.2, is now one of the most recent fields of LSP studies: the popularisation of science.

The important change in LSP research in the first years of the new millennium is certainly the step out of the special language limits. The communicative change in the 1980s emphasised the significance of crossing the textual border by taking pragmatic and contextual aspects into account. Now, the LSP scholars leave their habitat by looking at how science is mirrored in everyday domains such as TV shows or newspaper articles. In today's times of non-restricted global access to specific information and in today's times of facing the responsibilities for epidemics, such as HIV, informing the public at large has moved into the focus of LSP research and will remain one of the key issues in the next years.

2.1.2 What is special about special languages?

So far, I have used the term LSP without actually clarifying in what sense we are speaking of *special* language. Usually, special domains or special subjects are in our minds firmly associated with using a *jargon*, a professional language that deviates from the general norm. Hence, the notion of specialness here refers to professional fields and the respective experts communicating about special subjects within this field. Yet, LSP research investigated a variety of different levels that can represent specialness and characterise scientific or technical language. The communicative change in the 1980s made a plea for placing special language in a broader communicative setting:

> When we look at text as discourse, we look at its communicative dynamics: at the way in which various elements function in relation to one another to communicative patterns of integrated meaning. (Crombie 1985: 57)

The discussion of special languages now considers different interrelated elements that together form part of the communication process. In order to answer the question of what is special about special languages, it is useful to isolate the individual elements and illustrate the communicative process by using a model that pictures those basic factors.

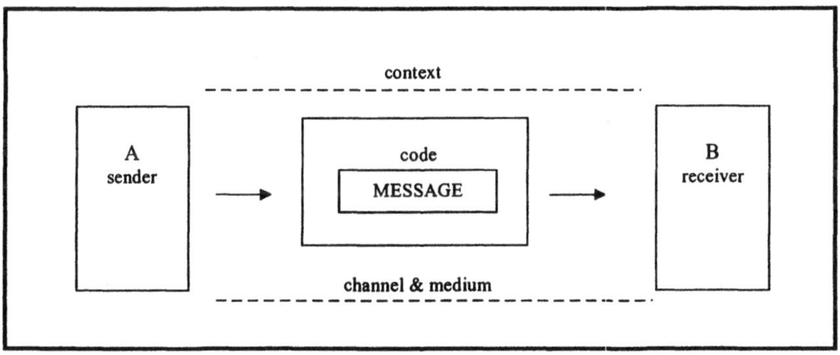

Figure 2: *Communication Model (adapted from Veestergard 1985)*

For this purpose, in Figure (2), I have modified a model taken from Veestergard (1985) that describes the fundamental elements involved in the communication process and sets them in relation. Broadly speaking, discourse takes place between at least two participants; one acts primarily as a sender (A) and the other primarily as a receiver (B). The general purpose of communication is to convey a message that fulfils a certain function, through a visual or an oral channel. Transmission via the visual channel can be realised by using, for example, a written medium such as a letter or a newspaper article. To convey the message from A to B, the sender converts a mental concept into the respective visual or phonetic representation and delivers this code via the chosen channel. If successful, the receiver decodes the message into the very thought the sender had in mind and recognises the intended function. If problems occur at any of the previous stages, the message cannot be decoded successfully and the attempted communication fails. As discourse is dynamic and highly interpersonally motivated, the success strongly depends on the people involved and the overall situation, the context, they are in. Thus, contextual factors influence the structure and the development of the discourse in advance. So, for taking the components of the communication model into account, I would argue that the concept of specialness is reflected at each level and in each respective element involved in the process. For that reason, specialness can be found in:

- context (domain, subject-matter, setting),
- discourse partners (expertise and status),
- message (specific functions),
- channel and medium (discourse path and text type),
- code (syntactical, morphological and lexical features).

According to Baumann (1992: 170), *context* in LSP research is defined as the total of situational conditions comprising professional, environmental as well as social premises. Professional conditions include first and foremost the framework of the domain itself, that is, for example, the domain of medicine, law, philosophy etc. and norms and conventions of the respective discipline. In a more narrow sense, it also considers the individual object of communication, together with the composition of the audience and their familiarity with the respective subject-matter. Environmental context refers to the actual physical setting and thus takes location and also time conditions into account. The physical setting is not only important for interpersonal communication, but can also play a major role in written discourse. Sager et al. (1980: 96) explain that the writer must not only be sensitive to the geographical area in which his text is to be distributed (e.g. in the UK or in the U.S.A), but also to the situation and location in which it is to be received (e.g. an emergency manual). Temporal factors include the time available for production and reception of the utterance and also consider potential options for repetitive reception or feedback.

Apart from the general setting, the individual context of the discourse partners plays an important role; this is what Baumann calls the *social context*. It considers the social status and role and relation between the individual communication partners. Special language communication is sometimes also called expert communication to refer to the notion of expertise that is attributed to members of a special discipline. Lothar Hoffmann even claims that special language does not exist without expertise: "Echte Fachsprache ist immer an den Fachmann gebunden" (Hoffmann 1976: 31). Expertise is achieved by gaining special knowledge and, as Fluck (1992: 221) claims, full expertise also means linguistic competence: "Unbestritten sind die Notwendigkeit von Fachwissen und die als Ideal anerkannte Einheit von Sprach- und Fachwissen [...]". Therefore, it is essential to take the individual context of the communication partner or audience into consideration. According to Adler and Rodman (2000: 111), this can include the interlocutor's status and self concept, prior knowledge, past experience and expectations and needs.

All of the above conditions have to be considered when choosing a *channel* and a *medium* for communication. The channel can be either oral or visual or even a combination of both. The oral channel, also called auditory channel (Sager et al. 1980), refers to spoken language, whereas the visual channel includes mainly written language, but also symbols, graphic representations or moving images. The medium of film would be an example for an oral-visual channel, similar to a talk based on written notes but read aloud. A medium is thus the physical representation with the help of which the message is passed on via the chosen channel. Special languages primarily relate to the written medium and in particular to special texts (*Fachtexte*). In her work on *Fachtextsorten des Englischen*, Gläser (1990) investigated comprehensively into text types in academic and scientific speech. Apart from defining them as a complex and coherent form of expression in a specific domain, she underlined their role as standard *Bildungsmuster*. They are consequently not only used as a means of convergence for the individual message, but furthermore constrain the structure of the message itself. The number of text types appears to be unlimited and to give a brief overview of major text types alone would go far beyond the scope of this study. I will thus comment on what is seen by text-oriented LSP linguists (e.g. Gläser, Baumann, Sager) as a prototypical form spanning a variety of disciplines, that is, the scientific essay or research article. It is seen as a prototypical written form which has a recognisable structure that separates the text into preliminaries, the main body of the text and completed by supplementaries. Nevertheless, the essay is conceived as an indivisible unit that gives the author's personal concern, presentation and interpretation of a scientific topic.

What is transmitted via the chosen channel and medium also depends on the message itself. The message lies at the heart of discourse as it transfers the utterance and its intended function to the audience. Selecting and organising information for a specific purpose and a specific audience also involves specific discourse functions. LSP is a restricted type of discourse, the primary goal of which is to spread information. For this purpose, it only needs a limited selection of functions. Trimble (1985) investigates what he calls the different *rhetorical functions* of scientific or professional communication and describes them as follows:

> I define a rhetorical function as a name for what a given piece of discourse is trying to do [...] and for the way in which the items of information chosen relate to one another or to the main subject of the given unit of discourse. (Trimble 1985: 12)

He identifies the five basic rhetorical functions of describing, defining, classifying, instructing and the visual-verbal relationship between a visual aid and the accompanying text. Nonetheless, one has to be aware of the fact that Trimble restricts these functions to a narrow setting, serving a limited purpose, for example, explaining an experiment or introducing a new theory or hypothesis. The more specialised a message is, the more specialised are the symbols we use to represent this special function. The message is transmitted to the audience by taking a certain form, i.e. a special written code.

As far as *code* is concerned, there are certain internal characteristics which typify, or must be essential components of a discourse, that make it recognisable as more or less scientific. The most prominent feature is undoubtedly the distinctiveness of the vocabulary. Yet, as Crystal (2003) explains in the following, it is important to not only look at terminology in isolation, but also to take its syntactical environment into account.

> It is possible to grasp a vocabulary of an area of scientific enquiry, yet still have a major difficulty in comprehension because of the way the sentences and discourse have been structured. (Crystal 2003: 272)

The syntax of scientific texts is shaped by the obligation to explain abstract phenomena in complex sentences incorporating a huge amount of specific information in one single segment. Beier (1980: 56) claims that technical language is characterised by "rather long sentences containing many clauses, often in complex degrees of dependency and with much embedding". To use space economically, the expert is supplied with bundles of relevant information only, using dense language and limiting the utterance to the most important, meaningful structure. Those structures lead to the nominalisation of scientific discourse as the dominating word class is formed by nouns, followed by adjectives and often drops the finite verb and is thus elliptical in nature. According to Biber et al. (1999), ellipsis is the omission of elements which are recoverable from the linguistic or situational context. This grammatical incompleteness of sentences is a salient feature of special language, since it does not affect the meaning of the utterance in a negative way to omit the finite verb.

According to Beier (1980), ESP texts are further characterised by a high frequency of present tense and passive voice used in the finite verb. Traditionally, tense refers to a change of verb form, indicating the time at which an action takes place. The present tense in LSP texts is used non-temporally to imply general truth of the statement. The non-temporal use supports factuality and universal validity which aims at presenting the finding as generally accepted and as proven and undoubtable facts. Passive voice additionally underlines factuality and also stresses impersonality and agentlessness. It refers to the importance of an abstract action or a finding as such, by omitting the person who carried out the action. Broadly speaking, the LSP text aims at creating an apparently neutral and objective scientific point of view. The latter is also supported by the absence of personal pronouns which leads to a rather impersonal style.

To sum up, linguists investigating LSP agree that special language has a more complex sentence structure than common language, making frequent use of long, complex sentences. Passive voice and present tense signify factuality and universal truth. Scientific texts are, for the same purpose, usually agentless and contain no or few personal pronouns.

Despite of all these distinctive syntactic structures, specific vocabulary is still regarded as lying at the heart of special language. Terminology marks a technical language and determines its form, distribution and usage. Terms are, according to Beier (1980: 31), symbols of mental concepts, serving to classify individual objects and processes. Thus, medical terminology, as the sum of individual terms, is the verbal expression of special subject matter and knowledge. It serves as a precise, concrete and unambiguous tool, enabling efficient internal communication between subject experts. Since lexis has always been of special significance in LSP research, I will now discuss it in detail in the following section.

2.1.3 The lexical level: terminology

As outlined above, the study of special languages in the past was predominantly concerned with the study of vocabulary. Linguists have always agreed on terminology of special languages being the most obvious distinguishing characteristic and folk belief even says that there are no special languages, but only special lexical terms. Crystal (2003: 372) confirms that the "first impressions of the language of science are that the distinctiveness lies in its lexicon" and Haage (1993: 238) adds that terminology is the most prominent, thus immediately eye-catching feature, and should be the primary object of discussion in LSP research. Analysing lexical properties of specialised discourse forms remains one of the key issues in today's approaches to special languages. This continuous linguistic interest in the lexical components of domain specific discourse is, as Niederhauser explains, due to their *signalling* function:

> Diese Fixierung auf den Fachwortschatz ist kein Zufall, sondern hat mit der wichtigen Rolle der Fachwörter in der fachlichen

> Kommunikation zu tun und vor allem mit ihrer Auffälligkeit zu tun. Fachwörter wirken als Signale der Fachlichkeit [...].
> (Niederhauser 1999: 26)

Hence, certain vocabulary items can act as signals, indicating specialness and scientificness. But what are the distinct properties and features of those terms that make them more or less scientific? In order to understand the reasons for the distinctiveness of terminology, it is essential to carefully examine its etymology and development over the past centuries and millennia. I will do this in the following by taking a closer look at the development of medical terms, not only because of the importance of medical terminology for the topic of this study, but also because of the prototypicality that is presented by the field of medicine.

Professional medical discourse with its high degree of classical terms is a prototypical language with a solid Greek and Roman foundation. The Greek school of philosophers produced men like *Pythagoras*, *Aesceulapius* and *Hippocrates* who placed medicine on a scientific footing and manifested their knowledge in form of the first written medical works. Today's pharmacology is based on the most important pharmaceutical writer of ancient times, the Greek *Pedanius Dioscurides*. He simplified the *pharmacopoeia* (index of pharmaceuticals) by selecting only proven, effective and non-superstitious remedies. After the Roman conquest, this precious knowledge was passed on to the Latin scholars and scientists, who translated the works into Latin. Written and scientific Latin survived the fall of the Roman Empire and was spread by mainly religious scholarship, through times and countries. Medicine was studied in the monasteries across Europe and numerous manuscripts (mainly translations) were handed down to religious and secular schools. Latin was the universal medium for scholarship and as Dirckx (1985: 43) states, "the knowledge of Latin became not only a key to the door of learning, but a means of communication among scholars across the length and breadth of the civilised world". This was true not only in medicine and theology but also in physical, mathematical and biological science and especially in jurisprudence; a great many modern legal terms are Latin. Christian missionaries also carried the first classical medical terms to Britain. Still, anatomic terms were often taken from the English vernacular, for example *arm*, *neck* or *skull*, and only where no appropriate or polite English term existed, either the Latin or the Greek equivalent was used (e.g. *anus*). Generally, scientists used non-English forms for those structures whose existence was not generally known until then (e.g. *ovary*). Names of common symptoms or diseases that have been discovered before the era of scientific medicine were usually English (e.g. *measles*, *sore*, *wound*). Latin or Greek terms were also used where the equivalent English term was not specific enough. According to Dirckx (1985: 25), native words like *numbness* had too broad a range of meaning as it covered *anaesthesia* and *paralysis* and *muscular weakness*. As a consequence, the English term *womb* was specified by the further distinction between uterus, abdomen, stomach and intestine. Discoveries in the eighteenth century led the way to further enlargement of the medical lexicon as new terms for new phenomena were based on Greek and Latin constituents. *Rubella* (*Röteln*), for example, was formed from the Latin word for red: *ruber*.

Nowadays, Latin and Greek are still predominant within the language of medicine and the classical languages are rich sources for stems and affixes, needed for the creation of new words which are nearly self-explanatory for educated experts. The use of Latin has, besides other benefits, the overall advantage that, as Michler (1981: 5) explains, the terms and words have a fixed meaning and nomenclatures are available to everyone, everywhere. Even new findings within the field of science do not cause a change in meaning. From this point of view, professional medical language with its high degree of classical terms can be seen as a language which provides constant semantic reference, easy access, and a compulsory nomenclature and is therefore exceptionally factual and productive. (Medical) terms are transparent as the meaning can be deduced by analysing the semantic properties of the morphemes that form the term. Niederhauser (1999: 32) states that transparency is essential: "[E]s ermöglicht jedem Wissenschaftler, die verwendeten Begriffe durch Rekonstruktion zu klären und den eigenen Sprachgebrauch explizit zu regeln". The prefix *sub-* for example signifies *under* and the suffix *-itis* refers to an inflammation. This is one reason why most technical terms referring to modern diagnosis and therapy are coinages built of Latin or Greek elements, which have a classic appearance but are actually modern, neo-Latin inventions. The formation of new terms follows classical patterns and is thus subject to strict rules. As a conclusion, the (classical) heritage of terminology outlined above leads to specific characteristics that are regarded as essential components of efficient domain specific discourse (cf. Hoffmann 1988, Fluck 1980, Sager et al. 1980). To sum up, the most important properties of terms are:

TRANSPARENCY
The meaning of the term is constructed by combining the semantic properties of its individual semes; e.g. *electrocardiogram* refers to the electrical (*electro-*) monitoring (*-gram*) of the heart (*-cardio-*). Thus, comprehension of expert knowledge is ensured by a systematic deduction of semantic information.

STANDARDISATION and INTERNATIONAL CODIFICATION
Terms and component semes are codified in international nomenclatures, in which the designates are ordered, classified and defined. Nomenclatures provide a fixed point of reference and guiding rules for interpretation and formation of new terms. They are thus an efficient tool for standardisation.

STABILITY
Terms are permanently defined and are not subject to change. This rule is strictly applied in the nomenclatures and has been raised to the status of law by Brown's *Law of Priority* in 1954.

PRECISION
A single concept is referred to by a single term or component and a term or component should represent only one concept and is thus monosemic. However, one has to keep in mind that this clear-cut reference and fixed definition holds true only in field specific use. The term *fever* for example is, according to *Psychremble's Clinical Dictionary*, only correctly applied when the body temperature rises above 38,5 degrees Celsius, whereas in folk use it is used for referring to any deviance from the normal body temperature.

NEUTRALITY
Latin and Greek designations are free from connotations as they provide concise information about e.g. physical conditions only. *Myocardial infarction*, for example, refers to the affected organ (*-cardial*), location (*myo-*) and type of event (*infarction*). By contrast, the lay term *heart attack* employs a metaphorical expression by using *attack*, which is taken from the source domain of warfare and thus conveys negative associations and emotions.

ECONOMY
Complex syntactic constructions are avoided by combining the respective components into one expression, for example, "a congenital absence of a normal orifice or tubula organ" is simply referred to as *atresia*.[3] This also leads to the nominalisation of scientific language style.

2.1.4 Horizontal and vertical division

Up to this point, I have employed the term special languages in a wider sense to exemplify individual use by referring to the case of medical language. In fact, an all-embracing paradigm of special language does not exist, but only particular domain-specific languages that contribute to a broad scheme. Beginning with early LSP research itself, continuous efforts to structure the various special fields contributed to an overall system of special language. From Wüster (1931), to Hoffmann (1985) and Roelcke (1999), linguists agreed on describing the phenomenon of LSP by integrating its components into a horizontal – and

[3] All medical definitions and explanations are either taken from *Dorlands Medical Dictionary* (30th Edition, 2003) or from *Pschyrembel* (260th Edition, 2004).

extended vertical – model. The resulting different levels of division suggest that domains exist side by side and are related to each other as they form part of a complex hierarchical framework. Figure (3) presents such a model of horizontality, aiming at illustrating the bonds between the individual disciplines rather than their diversification:

SPECIAL LANGUAGE				
Language of Science	Academic Language	Language of Technology	Language of Economics	Language of Consumption
Lang. of Natural Sciences	Lang. of Arts and Humanities	Lang. of Material Production	Lang. of Business Admin.	Lang. of Sales and Marketing
Lang. of Biology	Lang. of Linguistics	Lang. of Mechanics	Lang. of Accounting	Lang. of Advertising

Figure 3: *Levels of horizontal division (adapted from Roelcke 1999: 35)*

It goes without saying that such a model cannot claim to encompass the system of LSP in its entirety but can merely seek to present a fraction:

> Es kann uns nicht darauf ankommen, eine fertige und vollständige Gliederung der Fachsprachen vorzulegen. Praktisch ist dies nie zu erreichen […]. Es ist nur möglich, die untersuchten bzw. zu untersuchenden Subsprachen in einer offenen Reihe nebeneinander zu stellen. Allerdings kann sich dabei eine sinnvolle Platzierung – eine Art Skala – ergeben […]. (Hoffmann 1985: 58)

Thus, Figure (3) is first and foremost not to be interpreted top-down, according to the level of specialisation, but horizontally, according to the relation at a certain level of abstraction. In doing so, infinite rows can be constructed that are structured along a particular scale. Although the rows are theoretically endless, a scale has to have at least two (fictive) extremes that serve as points of orientation for the ranking of the individual elements. An obvious point of departure is provided by the languages of the "classical" scientific and academic disciplines, referring to long established fields of scientific discourse, such as the natural sciences or arts and humanities. They have a strong basis which is concerned with developing theories as well as with bringing them into a codified linguistic form that can be passed on for academic and educational purposes. Technical languages are characterised by stepping away from the theoretical approach towards a more practical implementation. Nevertheless, Roelcke (1999) explains that they lie at the heart of special languages, since technical languages possess typical characteristics, such as a highly specialised terminology and nominalised, formulaic language. Economic and business oriented languages are closer to the opposite extreme of the scale, that is, closer to the general language norm. Kalverkämper (1988) investigates the language of consumption (*Sprache der Konsumption*) and claims that its structure serves the communicative need of the public and is therefore the counterpart to restricted scientific language.

In addition to dividing special languages horizontally according to domains, the vertical division complements the model by adding a communicative and discourse-oriented dimension. Verticality can traditionally be divided into either two or three levels. The Prague school linguist Beneš (1969) was the first to distinguish between a theoretical and a practical style. The theoretical style is used for discussion on a scientific basis within a special field, whereas the practical style is used for communication on a professional basis within or outside the special field. This leads to a further division into internal and external discourse, depending on whether the communication takes place between expert and expert, or between expert and layman. Von Hahn (1980) adds an intermediating layer to Beneš's classification: the theoretical-practical style. He claims that theoretical language has to be used only at an abstract level, i.e. in research or education, and has to be seen as synonymous with scientific language (*Theoriesprache* = *Wissenschaftssprache*). Theoretical-practical language, or *fachliche Umgangsprache*, links theoretical knowledge with real situations as it applies theoretically gained information in everyday scientific work. Practical language is located at the lowest level of abstraction, distributing information among a heterogeneous audience. These traditional approaches are at the moment under discussion as the popularisation of science brings in an additional perspective. Information that was formerly restricted to the scientific community is now discussed by the public at large, leaving behind the scientific field. Therefore, I will revisit verticality in more detail in section 2.2.

In the following illustration (Figure 4), I have combined the two traditional concepts of horizontal and vertical division and related them to the particular field of medicine. Medical language within the context of LSP can be divided horizontally according to the subdisciplines and respective areas of expertise, as well as vertically according to the basic levels of communication.

Language of Natural Sciences

⇨

Language of Medicine

⇨

Theoretical disciplines ⇨	**Clinical disciplines** ⇨	**Vertical level**
Anatomy	Cardiology	⇩
Physiology	Immunology	
Biochemistry	Internal Medicine	internal (i.e. scientist & doctor)
Pathology	Neurology	external (i.e. doctor & patient)
Pharmacology	Paediatrics	

Figure 4: *Division of medical language*

As Figure (4) illustrates, language for medical purposes is embedded within the general framework of scientific language and can be horizontally divided into theoretical disciplines, which involve a high degree of research and usually little or no interaction with patients, and into clinical disciplines, which are concerned with diagnosis and treatment itself. According to Michler (1991), the clinical disciplines can be distinguished in relation to organic systems (e.g. cardiology = heart), characteristics of patients (e.g. paediatrics = children) and methodology (e.g. surgery). Vertically, medical discourse distinguishes primarily between internal communication on the one hand and external discourse on the other. Internal communication can take place on a scientific basis between a) scientist and scientist and b) scientist and doctor. The primary purpose of a) is to discuss research processes and findings, whereas b) additionally involves the approach to connect theory and application. Furthermore, internal discourse can take place on a practical basis between c) doctor and medical staff using a clinical language implemented in the routine of everyday work. The dominating text types of internal written communication include scientific essays and abstracts, guidelines, case reports and instructions. The external discourse is primarily patient-orientated and includes personal doctor-patient interaction as well as the addressing of an abstract group. The latter embraces forms of communication directed to an anonymous audience, sharing special features, for example suffering from the same disease. Those discourse varieties include educational campaigns, realised through television, billboards or brochures, as well as popular medical journalism and commercial information about specific treatments or products. As this study deals with aspects of science journalism and its popularisation, I will take a more detailed look at this topic in section 2.2.

2.1.5 Register, genre, discourse and text type

Before moving on to scientific discourse popularisation, I would like to briefly comment on the use of some ambiguous linguistic terminology. The use of the expressions *register*, *genre*, *discourse* and *text type* is treated differently by linguists and literary scholars who work in individual traditions and there seems to be particular confusion about the terms genre and register. They can certainly not be used interchangeably although they may overlap to some extent. Lee (2001: 46) claims that register and genre are two ways of looking at the same object: "Register is used when we view a text as language [...]. Genre is used when we view a text as a member of a category, a culturally recognised artefact." *Register* is, according to Halliday et al. (1964 and 1985) defined as "a configuration of meanings that are typically associated with a particular situational configuration of field, mode, and tenor" (Halliday et al. 1985: 39). Here, *field* refers to the subject-matter; *tenor* to communication partners and their relationships and intentions, whereas *mode* refers to the channel and medium chosen for the communication. Each of these elements is a variable itself that, if it is altered, leads to the construction of different registers. As a result, various types of register are available in a language system and register analysis assumes that the individual language choice is pre-determined by the communicative context. Medical English, within the framework of LSP, is an occupational variety or register, a distinctive language associated with profession. It allows personal expression with reference to a particular field and provides a set of language items associated with a particular situation or profession (tenor).

In referring to *genre* I am following Swales' canonical work on genre theory in academic and research settings. Swales (1990) was among the first to introduce genre analysis in LSP research. He emphasises that genre has to be seen as:

> [A] class of communicative events, the members of which share some set of communicative purposes. These purposes are recognized by the expert members of the parent discourse community, and thereby constitute the rationale for the genre. (Swales 1990: 58)

His fundamental conceptualisation has been interpreted by Barbara & Scott (1996) in so far as they include not only the notion of sharing a communicative event and purpose but also that of shared restrictions on form and content. According to Swales, this refers to the rationale of a genre which determines not only the schematic structure of the discourse, but also limits the choice of stylistic devices. He claims that those constraints are reflected within the individual *genre exhibits*. Genre exhibits include, first and foremost, the notion of text types. As already mentioned, Gläser (1990) conducted a comprehensive investigation of special text types, such as monographs, reports and – most importantly – research articles. The scientific research article is regarded as prototypical and standardised text type in LSP and was therefore closely examined by various scholars (cf. Bauman & Kalverkämper 1992, Gläser 1990, Swales 1990). Research articles and bulletins are predominantly published in scientific journals, for example, in the *Edinburgh Medical Journal* which is the oldest medical journal in English. First published in 1735, it claims to form the basis of "the changing epistemological norms of medical knowledge, the growth of a professional medical community, and the periodic redefinition of medicine vis-à-vis the non-medical sciences" (Atkinson 1992: 338).

There is even some confusion about the concepts of *text type* and *genre*. Stubbs (1996), for example, uses *genre* and *text type* synonymously, whereas Biber (1988) makes a clear distinction between both terms. He argues that *genre* is based on external criteria (e.g. audience, purpose, domain) whereas *text type* is based on internal characteristics of the text itself (e.g. lexico-grammatical properties). *Text types* are regarded as groupings of texts that show the same characteristics. However, Lee (2001: 39) clarifies that in Biber's sense two texts may belong to the same text type even if they come from two distinct genres.

The relation between *discourse* and *genre* is that genre places discourse in the communicative context (*register*) in which it occurs. Here, *discourse* relates, in accordance with Trimble (1985), to a set of language means (e.g. texts) that form a coherent and cohesive communication within a certain field. In the conception of *discourse community*, Swales (1990) combines all previous ideas. A *discourse community* shares a language with a common set of goals, common mechanisms of communication, common genre and a common but specific lexis. Finally, it should not go unmentioned that the term jargon, since it is often perceived as a loaded word. In a dictionary, it is neatly and neutrally defined as the technical vocabulary or idiom of a special activity or group. Yet, I agree with Crystal (1999: 174) that this sense is almost completely overshadowed by another "obscure and often pretentious language marked by a roundabout way of expression and use of long words".

2.2 The popularisation of science

Up to now, science popularisation has been a subject that has been studied only little from a linguistic perspective. It focuses on a form of communication which is all set to provide valuable future insights into the dynamics of language use in particular settings. The study of discourse popularisation is described as the *meeting point* between science and the general public (Moirand 1997) and addresses some crucial questions:

> We need to question who the actors are, how the various discourse levels interact, what modes are involved and what is communicated, and we need to consider what these questions imply for analysis. (Myers 2003: 267)

It is beyond the scope of this study to provide a comprehensive answer to all these questions. Instead I intend like to outline the various elements that are involved in popularisation by taking a closer look at the concept of verticality and thus at barriers to communication and at strategies that seek to overcome this barrier.

2.2.1 The concept of verticality

As discussed earlier, the state-of-the-art model of special language division is a two-dimensional one: the horizontal axis gives the relation of domains and their respective level of specialisation and the vertical axis presents particular communicative situations. The model mirrors variation and diversification in LSP discourse. Whereas the division at the horizontal level, as for example proposed by Hoffman (1985), is regarded as generally accepted and thus established, agreement on the vertical level is still out of reach. As already outlined in the previous section, Beneš (1969) introduces a stylistic dichotomy and distinguishes therefore between a theoretical and a practical style. A multi-layered form with three or more levels is favoured by von Hahn (1980) and Hoffmann (1985). They focus on different stages of theoretical and professional language use which are set apart by different communicative factors, such as the interlocutors. In this context, Hoffmann (1985: 65-66) defines five stages of specialisation or abstraction, each one associated with different language patterns:

- **Highest level of specialisation**: artificial and formulaic language (e.g. symbols, numbers, abbreviations), used in theoretical science (scientist – scientist);
- **High level of specialisation**: artificial language with some formulaic but also some natural elements (e.g. non-elliptical syntax), used in research (scientist – doctor);
- **Medium level of specialisation**: natural language with high degree of formulaic elements and terminology, used on a professional level between experts (doctor – doctor);
- **Low level of specialisation**: natural language with some formulaic and terminological elements, used on a professional basis between expert and staff (doctor – nurse);

- **Lowest level of specialisation**: natural language with isolated terminology, used for lay communication (doctor – patient).

Nevertheless, one has to be aware of the fact that theoretical or research-based special languages might show all of these levels but special languages that are more applied do not necessarily have to discuss a subject-matter at the highest level of abstraction.

The introduction of a vertical division created an awareness of the existence of a sort of external discourse. Traditional definitions of special language usually concentrate on internal communication only. That means that they put particular emphasis on expertise. The concept of verticality brought about by linguists like Hoffmann reflects this idea, because he and his colleagues still view communication from an internal perspective. This traditional standpoint has lately been criticised, because the popularisation of science has added a new point of view. The public at large now discusses information that used to be restricted to the scientific community. The information leaves the scientific field and the focus of LSP research started to shift from the upper to the lower level of the vertical scale:

> In the past, LSP research stressed internal special communication. The research object, for example, was often confined to communication between specialists. However, recently, external special communication has also been emphasized. (Schröder 1991: 15)

Since the early 1990s, the interest in a lay perspective on science has grown and the focus is no longer exclusively on the internal communication but also considers communication beyond scientific borders as a part of LSP research. The study of the popular presentation of science is today one of the key issues among linguists interested in academic or professional communication.

Hilgartner (1990: 528) regards popularisation as "a matter of degree, a continuum with two ends" and Busch (1994) claims that in simplest terms, the concept of verticality can be reduced to a dichotomy, to a twofold model with the expert and the lay perspective as extreme opposites. Busch challenges the traditional interpretation of internal and external communication as he argues that external discourse has to take place completely outside of the scientific field and is therefore conducted by laymen only. The communication formerly referred to as external, between expert and layman, is thus in fact transcendent or intermediating communication. In this respect, Ciaspuscio (2003) develops a model which assumes the existence of an elementary three-cornered circuit including scientists, the general public and mediators (e.g. journalists) and regards the spread of scientific knowledge as a one-way process. One way in so far as science is regarded as the source of information which is passed on to a mediator who then conveys the information to the public. This process emphasises the role of mediators, who are the link between the world of scientists and the world outside, that is the general public. Tom Radford, a journalist writing for *The Guardian*, directly addresses medical scientists and stresses the necessity to find a common ground in order to communicate without difficulties and misunderstandings:

> So we - journalists and scientist - are left with a big problem: how do we arrive at a language in which you can tell and we can listen? This is a necessary step, this is democracy's demand. (Radford 2001: 354)

The popularisation of scientific facts implies that science is built on an in-group and distinct language that needs some sort of decoding. This calls for a mediator, an agent who masters a special language to some extent and is able to translate it into everyday language. Weigold (2001: 171) explains that generally speaking the *communication link* from scientist to the public will travel through an information professional or a news organisation. Yet, one has to point out that journalists are not to be regarded as experts in the particular scientific fields since they have a certain insight but frequently lack the necessary educational background. Access to scientific discourse requires not only domain-specific education and practice but also awareness of the codes and norms used. Calsamiglia (2003: 141) argues that "the linguist usage we typically find here is conditioned by a special rhetoric that has to be learned". This means selecting ways of expressing and formulating within the strict limitations of the discipline or speciality in question.

The vertical division of medical discourse includes the use of specific language in research and clinical settings at the internal level. Intermediating communication can be either immediately patient-directed, as in a face-to-face consultation, or via a mediator. If the mediator for his/her part addresses the public and the public discusses the topic outside of the scientific borders, external communication will be the result. In his most recent work, Busch (2001) takes a closer look at communication in so called health-circles (*Gesundheitszirkeln*) and confirms that the lower level of vertical discourse in medical settings is manifold. He examines doctor-patient communication as well as discussion in self-help groups and also discussion of health topics on TV, in print media and on the internet. The discourse forms differ from each other but they have a common objective: to pass on information about medical conditions and diseases in a comprehensible manner.

Taking both traditional and modern perspectives into consideration, I suggest the following basic model of verticality (Table 2). It will serve as a platform when I examine the circulation of information about SARS.

Table 2: *Vertical division of popular scientific discourse*

Vertical level	Degree of Specialisation	Communication partners
internal	high	expert - expert
intermediating	medium	expert - layman
external	low	layman - layman

The model comprises three fundamental levels: Communication at the internal level belongs within the boundaries of a specific field and thus involves a high degree of specialisation. Such a high degree of specialisation or abstraction manifests itself also in formulaic and dense language use. At the intermediating level, the interlocutors are experts (*insiders*) and journalists or information professionals (*outsiders*) who leave the restricted field behind. The worlds of science and public meet and the language use shows a medium degree of abstraction with still rather more scientific than general features. External communication takes place outside the scientific field and is realised by intermediaries and the general public or by the public itself. The language has a low level of specialisation and the general elements clearly dominate the scientific features. In an ideal case, the communication partners are aware of the gradience that such a division implies. Yet, in a real-life situation, a problem may occur when both levels 'clash'. For example, scientists tend to perceive their communication partner as similar to themselves and have difficulty in estimating the level of their interlocutor's knowledge. In this way, barriers to communication are built.

2.2.2 Barriers to communication

Special languages carry or realise a particular discourse with the objective to facilitate precise and efficient communication about a particular subject-matter. As a result, scientists express themselves in what a lay audience perceives as an unknown and hence difficult language. According to Crystal (1999), the use of a special language in a non-specialist setting can easily create a barrier to communication:

> [W]hen the subject matter is one where we feel implicated, and think we have a right to know, and the speaker uses words which act as a barrier to our understanding, then we start to complain; and if we suspect that the obfuscation is deliberate policy, we unreservedly

> condemn, labelling it gobbledegook and calling down public derision upon it. (Crystal 1999: 174)

Labelling scientific language *gobbledegook* means that, in simple terms, the specific features of scientific language – those that are beneficial within professional communication – are perceived as alien elements in external communication. They are not understood by the lay audience as they are unfamiliar and semantically non-transparent. As outlined before, a special language is not only characterised by its terminology alone, but also by its lexical density, impersonal style and complex syntactic structures. Hence, special language use in professional settings enables personal expression with reference to a particular field and provides a set of language items associated with the profession. The distinctive language use serves as an internal code and signals a particular group membership. By using a specialised medical language, scientists, researchers and physicians mark themselves as members of a particular professional *in-group*:

> Nicht zuletzt durch ihre besondere Sprache setzen sich die Experten von den anderen, den Nicht-Eingeweihten, ab. In der Expertensprache liegt ihre Macht; sie sitzen an den Schalthebeln, die nur sie bedienen können, in dieser Sprache und dank dieser Sprache können sie über andere verfügen. Gleichzeitig liegt darin aber auch ihre Ohnmacht. Je ausschließlicher sie sich der Expertensprache verschreiben, umso weiter entfernen sie sich von der allgemeinen Sprache, von der für alle möglichen und für alle verbindlichen Kommunikation.
> (Bausinger 1984 as cited in: Fluck 1991: 39)

Bausinger claims that approaching the *general language paradigm* is a crucial factor for bridging the language barrier. General language (*Allgemeinsprache*) and scientific language (*Fachsprache*) are not fundamental opposites, but are related constituents of one system with various connections and interrelations. A fixed definition for general language does not exist and not even a fixed term, as it is also called standard language, common language or every-day talk. I will use the term *general language*, considering the feature of universality referring to a potential language shared by all participants of a speech community.

Wichter (1994) argues that the process of popularisation need not negotiate between scientific and general language only, but must also bridge a knowledge gap. He underlines that conflicts are not only due to the language itself but also due to a knowledge asymmetry originating from the divergent backgrounds of the communication partners. This means that when experts and laymen communicate, a barrier may be created by the characteristic features of scientific language and also by the different levels of knowledge and expertise. As I have already pointed out, definite boundaries between science and non-science do not exist. There is an asymmetry of knowledge and a divergent linguistic structure:

> The objective of special language communication can therefore best be described as affecting the special knowledge structure by means of technique not available in general communication.
> (Sager et al. 1980: 69)

Owing to their own special knowledge and familiarity with the respective techniques, experts are not always aware of a layperson's communicative needs. Their profound domain knowledge leads professionals to rely strongly on their own semantic concepts when planning or producing information for laymen. If experts are not sensitive to the layman's lack of knowledge, the latter will be incapable of understanding the meaning of the message, since he/she cannot draw on the same education and past experiences. The terminology used by medical professionals refers to concepts that have no obvious equivalents in every-day life so that patients are unable to connect abstract concepts with a familiar representation. A contextual asymmetry frequently leads to miscommunication, creates communication barriers and may even result in an apparent unease about speaking to a scientist or medical professional:

> Den Patienten ist – Folge der Assymetrie – in aller Regel eine adäquate Kontrolle oder Bewertung ärztlicher Maßnahmen nicht möglich, vielmehr offenbart der Patient dem Arzt seine subjektiv sinnlich erfahrene Symptome und erwartet, der Arzt möge aufgrund seiner Deutungskompetenz eine angemessene Therapie einleiten. (Busch 1994: 40)

In other words, a process of mediation (*Vermittlungsprozess*) must pay attention to contextual factors, such as setting, participants, their identities and roles, as well as their aims, needs and previous knowledge. The popularisation of science has made various attempts to bridge or lower the language barriers. These approaches try to take the interlocutor's perspective into consideration and to be sensitive towards a layperson's *communicative competence*. Language can create communicative barriers but it can also be a bridge between science and public.

The concept of *communicative competence* was introduced by Hymes in 1972. He claims that, in order to communicate effectively, speakers are not only expected to show grammatical competence, but also to use language which is appropriate to the situational context. Whether an utterance can be regarded as appropriate depends on the context in which it is used and evaluated. Thus, communicative competence is the ability to use language in an adequate way by adapting to particular circumstances and also taking the intention and function of the utterance into account. According to Canale and Swain (1980), communicative competence can be broken down into four levels:

- grammatical competence: formal possibility;
- sociolinguistic components: appropriateness in regard to context;
- discourse components: appropriateness in regard to cohesion and
- coherence;
- strategic components: appropriateness in regard to the use of
- communication strategies.

In order to communicate efficiently, scientists have to have various competences to be able to adapt to different settings and communication partners. In an ideal case, they combine the knowledge of basic grammatical principles with the knowledge of how language is used and how discourse is structured in both the scientific and the public community. Consequently, competence is a dynamic process which depends on the negotiation of roles and not just linguistic proficiency alone. Analysing language in various authentic contexts should be one of the key issues in studying scientific discourse popularisation.

2.2.3 The vulgarisation of science

At present, there is an increasing interest in bridging the gap between the scientific community and the general public. Organisations like the *U.S. National Science Foundation* called a special *Science Popularization Division* into life, aiming at fostering the public understanding of scientific and technological matters. Yet, interest in science popularisation as an object of inquiry has grown for the most part within the last 20 years. Only a limited number of studies have investigated the patterns of communication outside the professional field. As Cooter and Pumfrey (1994) claim, popularisation is still regarded as a fairly recent tendency, especially in LSP studies:

> Surprisingly little has been written on science in popular culture, past or present. Still shrouded in obscurity are the effects of even the most obvious mechanisms for the transmission of scientific knowledge. (Cooter/Pumfrey 1994: 237)

The existing approaches to science popularisation include mainly sociological (e.g. Stichweh 1984; Karpf 1988) and linguistic studies which are primarily genre and discourse-oriented (e.g. Adams-Smith 1987; Nwogu 1991; Niederhauser 1999). According to the sociologist Stichweh (1984), there is an observable circular movement in the historical development of science popularisation. In the eighteenth century, a popular discussion of aspects of the pre-modern system of science began. Then, in the nineteenth century, the discussion came to a standstill when the classical scientific disciplines were further differentiated. Science was still open to public discussion but only in particular scientific clubs and public lectures. In the twentieth century, changing forms of communication led again to the opening of scientific discourse for the general public. The *professionalisation* in the nineteenth and twentieth century went hand in hand with scientific growth and had many valuable effects to expert communication. Nonetheless, it also promoted public unease about science.

> At the mid-20th century, what we find is that there was a traditional view according to which scientific production had very restricted circulation. Confined within the ivy-clad walls of institutions - the universities and research centres - articulated mainly in writing, the register used is far removed from what was normally comprehensible to the general public. (Calsamiglia 2003: 140)

The term *popularisation* is favoured over the earlier term *vulgarisation* (derived from Latin: *divulgare*) since the latter also carries a pejorative meaning. Jakob (1998: 711) speaks not only of *vulgarisation* but also of *trivialisation* of scientific language, because once the scientific context is left behind, the special language is not only simplified but quite often distorted and dumbed down. Weigold (2001: 175) even argues that science messages are often seen as being wasted when passed on to the general public. He describes the popular belief about *science illiterateness*, because "large numbers of American adults appear to be scientifically illiterate, leaving many to conclude there is a problem in science communication". He demands an improvement in scientific journalism and questions the necessity of providing the general public with insights into scientific findings. Moreover, he doubts whether it is useful for the public to understand, for example, the methodology of science. Russel (1986: 83-83), on the other hand, complains that for scientists communication with a lay audience is "almost always a secondary issue" and she blames scientists for not adapting to the situation but "talking in the most technical language possible".

Stichweh (1984) summarises the outlined concerns and concludes that popularisation cannot be seen as a simple, unitary phenomenon but involves multiple publics and thus multiple modes of popularisation. He identifies four basic modes: *interdisciplinary*, *pedagogical*, *political* and *general*. *Interdisciplinary* refers to communication with colleagues from neighbouring academic or professional disciplines who have a basic understanding of science, but not of the particular subject matter. *Pedagogical* popularisation is used for educational purposes, embracing the essential process of passing on the subject knowledge as well as the linguistic competence required for specific communication. *Political* popularisation is a phenomenon which occurs when scientific projects are presented to public institutions, for example to receive funding. The *general* mode of popularisation is the non-restricted which is in place when the public at large is addressed without the communicator's knowledge of the individual composition and characteristics of the audience. It is the general mode of popularisation that this study is concerned with.

Taking all of the previous arguments into account, it is obvious that bridging the gap between science and the public requires a shift in *audience design*. Particularly relevant in this context is Bell's canonical work on *audience design* (1984, revisited 2001) in which he relates the concept of audience design to speech accommodation theory introduced by Giles (1973a; 1973b). Accommodation means that we adjust our linguistic performance towards the style used by our communication partner in order to match verbal behaviour. With the introduction of audience design, Bell sketches out a model that, as a result, centres around the most important factors influencing language style Figure 5).

Bell sets shifts in style in relation to the process of adapting utterances to the communication partner's perspective. He claims that "variation on the style dimension within the speech of a single speaker derives from and echoes the variation which exists between speakers on the social dimension" (Bell 1984: 159). The theory of audience design assumes that speakers design their style primarily for – and in response to – their interlocutor. *Response* is regarded as the primary mode of style shift and it can be applied to all codes and repertoires within a speech community. In the case of scientists predominantly addressing members of the same speech community, the audience follows an ideal as they share the preferred mode and thus also form of communication within the restricted field. A special problem of action occurs when scientists have to adjust their verbal behaviour to an audience outside their special field that is to a lay audience that is familiar neither with conventions nor with forms of special in-group communication. Bell furthermore introduces the additional variables of *topic* and *setting*, suggesting that speakers can also shift their style because of a change in context and he complements the model with the *initiative dimension*, meaning that here a change in the situation is initiated by the style shift itself.

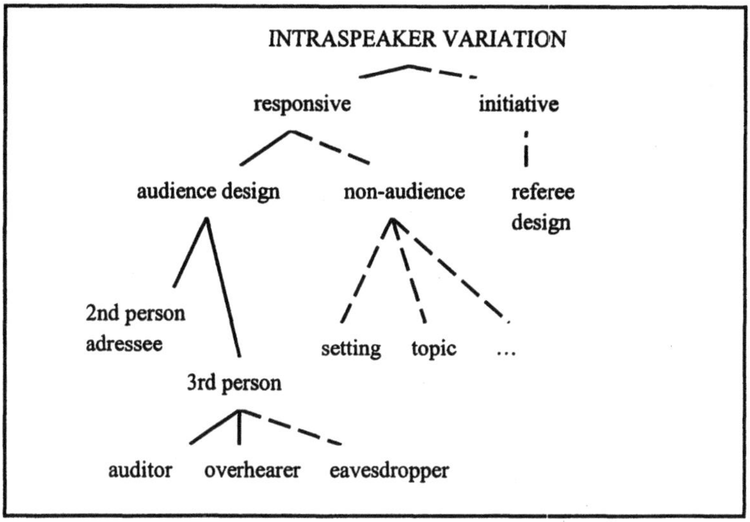

Figure 5: *Style as audience design (Bell 1984: 162)*

The study of popular scientific discourse is not primarily concerned with the categorisation of expressions into scientific and non-scientific, but with taking hold of stages and aspects in the course of the process and describing them adequately. It is therefore interested in linguistic means that are applied strategically in order to make scientific discourse comprehensible and thus accessible to a general audience.

2.2.4 Strategies of popularisation

A *strategy* describes a planned course of action to reach a certain goal. The main goal in discourse popularisation is to make scientific information accessible to a larger public. This means the choice of language has to vary according to the level of popularisation. Special vocabulary remains the greatest obstacle when scientific discourse is made more transparent and therefore receives a lot of attention. Ciapuscio (2003: 208) warns that a specialised lexicon is the most frequent cause for a "communicative conflict". Radford (2001: 354) supports this view by arguing that words in this sense are "the first investment". As the main barrier to understanding, the lexical inventory serves as a point of departure for a comprehensive analysis of popular scientific discourse. Apart from the lexicon, the complex syntactic structures establish further communicative barriers. For this reason, not only the lexicon has to be modified to meet the needs of the target audience, but also the complex syntactic structures (cf. Nwogu 1991).

Myers (2003) reviews current approaches to the popularisation of scientific discourse and claims that there is a *dominant view* which has to be discussed critically. The dominant view of science popularisation identifies a relation between the two opposing genres, but regards it as a one-way process of simplification, one in which scientific articles are the origins of knowledge that is then debased by mere simplification for the public. Simplifying the source text means – for the most part – changing the amount and density of scientific information (cf. Niederhauser 1999). In order to reduce the amount of information, unnecessary details are left out and only the most relevant points are given. The journalist Radford addresses scientists in a straight forward manner and makes a plea for using plain language to communicate efficiently with a lay audience:

> There are some simple rules for this. The language should focus on the core of science, rather than its details and it should be as of-the-people as possible. There are terms that scientists use all the time - *organelle*, *cytoplasm*, *isostasy*, *polymerase chain reaction* and so on - which have no real meaning for anyone else. So must you use them? And if you must, confine yourself to one of these words as people are prepared to struggle with one unfamiliar concept in a sentence, but not with two. (Radford 2001: 254)

The goal of popular science journalism is to communicate complex scientific information in a familiar and easily accessible way. This can be reached by reducing the amount of information and by achieving the correct balance between abstract and concrete points. The strategy of simplification also affects the text itself. Crystal (1998: 372) argues that the aspect of making text "visually manageable" should not be neglected. Visual simplification suggests the generation of smaller paragraphs with few sentences, and an uncomplicated noun structure including a low lexical density. To reduce the density, scientific information that is left within the text is, for example, followed by supporting explanations. Popularisation not only means omitting certain concepts, but it also involves the use of mechanisms and concepts that corresponds to the layman's prior knowledge and perception of the world:

> Es gilt, die in der fachinternen Kommunikation verdichtet und verknappt dargestellten wissenschaftliche Information in der fachexternen Kommunikation zu entfalten, umzuschreiben und mit dem nötigen Kontext zu versehen. (Niederhauser 1999: 128)

Hence, popularisation attempts to present relevant information in an appropriate way. It is far more than merely simplifying information and summing it up. Rewording abstract information in a non-abstract manner is one of the core processes in popular scientific discourse. However, that does not mean creating simple and transparent terms by merely replacing Latin or Greek-based terms. The general public is usually not able to follow the paths of science until it gets to know the basic glossary and terminology of a relevant subject. Specific terms, such as *coronavirus*, are not to be seen as a barrier to communication that cannot be overcome. If a term is essential for the development of a particular topic, it has to be commented on and has to be introduced to the reader (a *virus-family* that causes *flu*). If this is done carefully, this can even lead to a process called *determinologisation*. Becker (2001: 12) defines it as: "Gegenbegriff zur Terminologisierung, den Prozess der Rückübersetzung von Fachbegriffen in die Allgemeinsprache".[4] So the antonym to terminologisation denotes the process of back-formation, in which previously restricted terms enter the general lexicon and are thus institutionalised in every-day language use.

At the lexical level, popularisation not only involves determinologisation but also *reformulation* and *recontextualisation* of scientific discourse. This requires popular discourse to adapt to the conditions and constraints of the audience to which they are directed. Gülich (2003) is interested in those formulating activities that are particular to knowledge transfer, i.e. reformulation and illustration:

> Reformulations are particularly helpful in that they enable technical and non-technical terms to be used side by side is such a way that they illuminate each other. The main function of illustration procedures, especially of examples, scenarios, and concretisation, seems to be relating abstract medical information to everyday life and experience. (Gülich 2003: 258)

Producing a popular scientific text basically means recontextualising and first and foremost reformulating the source in such a way that it is comprehensible and relevant to a different kind of audience. This can be done by employing different lexical and semantic strategies. In an survey of linguistic studies concerned with this process, I have come across the following individual techniques:

- **The use of substitutional devices** (cf. Brand 2002, Busch 1994): A term is replaced by a familiar word of vernacular origin (synonym) that has

[4] In her dissertation on *Populärwissenschaftliche Vermittlungstexte*, Andrea Becker (2001) focuses on the explanation of scientific terms, the use of mental models like metaphors and repeatedly on determinologisation.

more or less the same meaning. Example: epistaxis is replaced by nosebleed.

- **The process of reformulating or paraphrasing** (cf. Ciapuscio 2003, Gülich/Kotschi 1987): A term is replaced by a discourse fragment that has more or less the same meaning. Example: lipostrophy is replaced by changes in fat tissue.
- **The use of definitions** (cf. Candel 1994, 1999): A term is supplemented by a brief description of some general and specific properties of the referent. Example: ribosome, the protein-making part of the cell.
- **The use of an analogy or a metaphor** (cf. Becker 2001, Rothbart 1997): A term is compared with an object that the reader is familiar with or it is replaced with a metaphor that is easier to understand. Example: virus is pictured as bug.
- **The process of generalisation** (cf. Gülich 2003): A term is introduced with the help of a proposition that extends the meaning from an individual to a universal concept (i.e. by using a hyperonym). Example: staphylococcus is expressed as bacteria.
- **The process of exemplification and concretisation** (cf. Gülich 2003): A term is introduced with the help of one or more propositions that exemplify the meaning from a general to an individual concept that exists in ordinary life. Example: synthesised epinephrine such as the hormone adrenalin:

The main challenge in the implementation of scientific terms in the general language is the choice of lexical devices that match and complement the competence and needs of the audience. In simple terms, a genre shift from a specific to a general audience requires a decrease in specialisation and an increase in comprehensibility. As depicted in Figure (6), this is also valid for producing popular scientific texts.

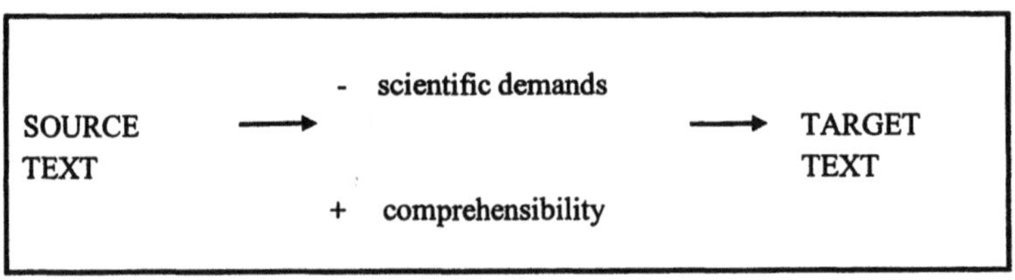

Figure 6: *Popularisation process: a simplified model*

Despite all previous efforts to describe the relationship between scientific and popular discourse, there is a recent debate that regards scientists and the general public as incompatible actors in separate worlds who are constrained in their use of discursive devices. In calling for a new reference point, there is a need to redefine the coexistence of scientific research publication and the news media. In his most recent article on this topic, Myers (2003) makes a plea for regarding the two discourse varieties as separate forms, with the popular newspaper version as stories in their own right. By doing so, he explicitly reproaches the dominant view that assumes that the information is transmitted from one discourse to another:

> This article critiques the dominant view of the popularisation of science that takes it as a one-way process of simplification, one in which scientific articles are the originals of knowledge that is then debased by translation for a public that is ignorant of such matters, a blank state. (Myers 2003: 265)

There is a growing uncertainty as far as the science-society interrelation is concerned. Gledhill (2000) questions the one-way transfer of information and he argues that information transfer consists of a complex network of structures which involves interaction between scientists and intermediaries, such as journalists and the general public. Logan (2001) supports this view and suggests an interactive science model in which the flow of science knowledge is not always from expert to layman but implies that it might be shared or multidirectional. Yet, Ciapuscio (2003) argues that from a production stance, it is obvious that popular science texts originate in other texts since these original texts (scientific or research articles) are produced to present and circulate scientific findings.

In this study, I intend to question the boundaries of scientific discourse and genres of popularisation: Is there a strictly linear process and a vast gulf between the two forms or do both discourses interact? What modes are involved on the lexical level and what is communicated in what way, e.g. by using which devices? Myers (2003) thinks that there is no evidence to be gained from a contrastive analysis of scientific and popular texts as they are not related. Keeping Bell's (1984) model of audience design in mind, I would argue that there are at least two independent dimensions in which they are related, namely topic and setting. In the case of the SARS-crisis, the topic constitutes the framework in which both "worlds" are united. Events, implications and findings are discussed on different levels but yet simultaneously. As a result, the dual realisation of one and the same restricted topic must show similarities as well as deviances.

2.2.5 Science journalism: news as discourse

In analysing the popular presentations of the SARS-crisis, the characteristics of popular scientific discourse coverage in the press must be sought in context. This section discusses – in brief – the mechanisms of scientific news journalism and the overall structure of news stories. Calsamiglia points out that in today's times of global mass communication, newspapers, television and the internet are the most influential sources of information and hence of science-related news:

> In recent decades, society's news media, responsible as they are for information in all areas, have taken it upon themselves to provide information with ever increasing intensity on scientific progress. As a result, at present, the social representation of scientific knowledge is actually derived in large part from the news media.
> (Calsamiglia 2003: 39)

In other words, the *newsworthiness* of science is linked to a concrete event rather than the idea of research itself. What is important is its application, its utility and its relevance to people's lives. The public wants to be informed about topics that have an immediate practical implication and about possible risks that might affect their lives. The coverage of the SARS crisis was of major public interest, because the epidemic presented a possible threat for people all over the world. Media coverage reflects public concern in so far as it focuses on potential risks, threats and precautions. When it comes to identifying threats, avoiding risks and taking countermeasures, the public takes a particular great interest. It is quite obvious that the media landscape reinforces this tendency by a broad coverage of the topic so that newspaper sales will go up. The saying "bad news is good news" denotes this negative aspect of journalism. Researchers studying science popularisation have repeatedly blamed the notion of sensationalism that goes along with reporting about public dangers and hazards (cf. MacDonald 1995). Journalists highlight the potential for a dramatic turn and tend to focus on strategies of personalising information rather than keeping it neutral. Radford defends this point of view:

> That may be a manifestation of the urge to shoot the messenger who brings bad news, or it may be a fair description. Journalism works by provoking, by titillating, by pleasing, delighting or shocking. People often talk about 'sensational journalism' but if by that they mean that journalists, like writers, use mechanisms to appeal to a sense of humour, a sense of delight, to a sense of horror, a sense of surprise, then they are not saying very much. (Radford 2001: 353)

The British journalist Highfield (2000) is convinced that scientists could even learn from journalists. Journalists think carefully about their audiences and choose their words accordingly. When reporters write news articles on scientific topics, they think of possible ways to connect scientific reality to everyday life. Nonetheless, one has to be aware of the opposing textual contexts. In contrast to news stories, research articles present and discuss scientific findings and are essentially bound to theoretical methods, conditions and perspectives. For this reason, scientists have to produce very detailed texts with the help of scientific terminology and formulaic language.

News reports are also bound to a certain structure that is partly determined by the complex production process. As outlined previously, journalists serve as mediators in scientific discourse popularisation. In this role, journalists have to interpret and to understand a particular topic before they can present the key aspects in an article which may refer to sources and include personal comments. The process of understanding is followed by the question of how to organise the pieces of information to meet the needs of the target audience. Van Dijk (1988) develops a model (Figure 7) describes the *superstructure* of news reports schematically and takes the tasks of a journalist into account.

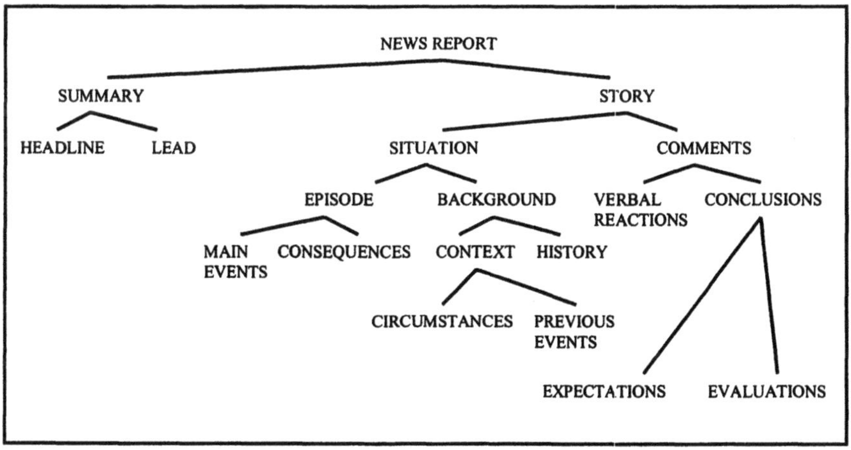

Figure 7: *Superstructure of news reports (van Dijk 1988: 55)*

The model assumes that general patterns recur in the design and construction of a news report and determine its paradigmatic and syntagmatic structure. At first, the schema distinguishes between two basic components, summary and story.

The summary includes the headline and the lead. They have an immediate eye-catching and hence attention-getting function and are designed to be distinct from the rest of the text. In order to catch the reader's attention, they summarise the topic of the article in precise and *snappy* words. The story is shaped by situation and comments. The situation includes the episode itself, that is the main event, as well as background information which complements the present affairs. The situation triggers verbal reactions, for example, comments from different sources. By adding relevant verbal reactions, journalists try to obtain a higher credibility for the facts they are presenting. They leave the readers enough room to develop an opinion that does not have to correspond to their own point of view since the manner of reporting is objective to some degree as the report originates from credible sources. By including personal comments, the journalist gives his/her own opinion and evaluation of the topic. The episode includes the main event and possible future consequences. The latter are decisive for the reader, since they may affect his/her life. Therefore, the newsworthiness of a scientific event depends on whether the event may have an immediate influence on the audience or not. Since words become the core of information, the textual structure should be kept as manageable as possible, providing the reader with small pieces of information, embedded in simple sentences, presented in an easy-to-follow way. Van Dijk stresses that the style should be adapted in order to create familiarity. This could be done by adopting a more narrative style or by avoiding passive constructions and thus personalising the text.

2.3 Functional perspective and research questions

Before (re-)framing the research questions the study seeks to answer, I would like to comment briefly on a perspective that might help to support observations and assumptions that are part of the theoretical framework of this study, namely the relation between form and function and the concept of functional dialects.

The fusion of a systemic view of language with a functional one has already been proposed by the Prague School linguists, also known as the Prague Circle.[5] Havránek, Jakobson, Mathesius and Beneš (to name but a few important members), approached the functions of language from the perspective of how these functions and their modes of realisation could influence the phonological, syntactic and lexical structure of a language. Jakobson expands the functional view on language to the issue of language development, primarily into the area of phonetics and phonology. He regards language as a tool and demands "to analyse all the instrumentalities of language from the standpoint of the tasks

[5] For a comprehensive overview of the Prague Linguistic Circle, see the *Prague School Reader in Linguistics* provided by Vachek (1964).

they perform" (Jakobson 1963: 482). Even earlier, Mathesius (1936) underlines the role of language as a means of communication and expression:

> In each case the starting point of the investigation will be the communicative needs of the speaker, and from this fact two consequences will of necessity follow; the way will lead from speech, as something which is immediately given, to language, as a system having an ideal reality only, and from the functional necessities to the formal means by which they are satisfied. (Mathesius 1926 as cited in Daneš 1964: 229)

Thus, the functional perspective contrasts with a more static view of language as it is seen as a dynamic whole of systems which varies according to intention, purpose or audience. Newmeyer (2000) speaks in this context of a form and function dependency in which the communicative situation motivates, constrains, explains and determines the structure of an utterance. This has a direct implication for the view on popularisation, as language has to perform a different function in different settings. This is closely connected to the idea of *functional dialects*. The Prague school linguists also introduced this notion of *functional variation* within language use that makes a distinction between prosaic and poetic language. This is, as Mukherjee (2005) describes, relevant to LSP as it can be regarded as a repertoire of linguistic means for specific purposes:

> More specific functional dialects may, of course, be identified; for example, the scientific dialect as a subclass of prosaic language, which is characterized by what is called the 'intellectualization of language'—lexicon, syntax, and reference conform to the overall communicative function that requires exact and abstract statements. (Mukherjee 2005: 1044)

In the context of functional linguistics, the Prague School linguists (cf. Havránek 1964) established the concept of *foregrounded item*. The idea of foregroundedness draws on the prominence of, for example, a certain item of vocabulary in a particular discourse that acts as a barrier to the *automatic* processing of an utterance. This view assumes that some parts of an utterance have a different effect on different readers, as these parts are specially patterned and thus linguistically deviant from the expected or familiar norm. As Niederhauser (1999: 141) indicates, medical terms can act as signal words and are perceived to some extent as forgegrounded items by a non-specialist audience: "In der Regel werden Verstehenslücken aber an den Fachwörtern festgemacht. Denn diese werden als Fachwörter erkannt und wirken gleichzeitig als Unwissenheitssignal." As a conclusion, medical terminology in popular texts consists of a special pattern, realised by a special form that could mark it as foregrounded.

The theory outlined so far leads to a variety of related research questions that have to be specified in order to set the stage for the following corpus analysis. It is useful to structure and classify the questions into the following three groups:

I The research questions that result from the different views on the popularisation of science are:

- Is there a relation between scientific and popular discourse?
- If there is one, is there a linear process of transformation?
- If there is none, how is the realised individually?
- Which variables are involved?

II The research questions that result from the focus on lexis, supplemented by the notion of foregrounding, are:

- Is there a different lexical realisation of a scientific topic in the two genres?
- If there is one, how does the lexis differ?
- Are there lexical patterns or structure that correlate?

III The research questions that result from the 'form-follows-function' assumptions are:

- How does a (specific) lexical item function in the individual setting?
- How do they intersect with a set of items in other settings?
- Do they create a common ground between the genres?
- how do they function in the complex interplay of science and media?

These questions constitute the starting point for a comprehensive analytical investigation of the lexical realisations in scientific discourse popularisation, based on a large corpus that represents the two levels of the SARS-crisis coverage.

3 Corpus-linguistic analysis

The obvious point of departure for a corpus-linguistic analysis is the data itself. Corpus compilation, corpus design and the question about how to approach the resulting amount of data need to be considered carefully to come to valuable conclusions. In this chapter, I will clarify corpus issues relevant for the analysis of the collected data, discuss requirements for corpus compilation, explain corpus characteristics and the methodology used. Finally, I will conduct a pilot study to put the corpus as well as the method to a first test.

3.1 Methodology I: Corpus theory and application

The first question that has to be answered before discussing methodological corpus-linguistic issues is what a corpus is and what it is not. Bowker and Pearson (2002: 9) describe a corpus as "a large collection of texts that have been gathered in an electronic form according to specific criteria". Francis (1992: 17), however, points out that in fact "many important corpora of English were assembled long before the computer was invented". McEnery and Wilson (1996: 177) propose a three-levelled definition that defines corpora in a loose sense as "any body of text", in the widest sense as "a body of machine-readable text" and in a stricter sense as "a finite collection of machine-readable text sampled to be maximally representative of a language or variety". For the purposes of this study, I prefer to follow a narrow definition of corpora and adopt the definition by Tognini Bonelli which comprises the most central elements:

> A corpus is taken to be a computerized version of authentic texts, amenable to automatic or semi-automatic processing or analysis. The texts are selected according to explicit criteria in order to capture the regularities of a language, a language-variety or a sub-language. (Tognini Bonelli 2001: 55)

Tognini Bonelli stresses the need for using authentic data and takes whole texts as the starting point. She furthermore refers to essential characteristics of modern corpora, such as machine-readability and computer-assisted analysis. Information about regularities or distributional patterns of particular sub-languages (or genres) can be extracted and interpreted. The explicit criteria that are required for the selection of the data will be discussed in the course of this chapter. Before doing so, I would like to comment briefly on the use of the term *corpus linguistics*. Mukherjee (2004) states in this context:

> In the light of the increasing heterogeneity in corpus-linguistic theory and methodology, one of the most hotly disputed issues at present is the question as to whether corpus linguistics should be seen merely as a new method (i.e. with virtually no theoretical implications) or as a new linguistic discipline or paradigm (e.g. in terms of an underlying corpus-linguistic theory of – and approach to – language.
> (Mukherjee 2004: 112)

He argues that it can be viewed as both, since it represents a new method and a theoretical approach in terms of *computer-aided descriptive linguistics*. So, the emphasis can be either put on the term *computer-aided*, that is on the methodological dimension, or on the term *description*, which would point towards a discipline.

As far as methodology is concerned, Jan Aarts (2002: 3) emphasises that "any discussion of corpus linguistic methodology must start from the data". A question that arises from this line of thought is how dominant the status of the data should be. Methodology-wise, a corpus can be used in different ways in order to reach results; either by seeking evidence for previously existing assumptions and thus searching the corpus systematically, or by exploring the corpus and developing assumptions while interpreting the data. In this context, Jan Aarts (2002) and Tognini Bonelli (2001) distinguish between the *corpus-based* and the *corpus-driven* approach. Tognini Bonelli explains *corpus-based* as follows:

> For the purpose of making a methodological distinction, the term *corpus-based* is used to refer to a methodology that avails itself of the corpus mainly to expound, test or exemplify theories and descriptions that were formulated before large corpora became available to inform language study. (Tognini Bonelli 2001: 65)

The corpus-based approach can be regarded as a method in which the researcher uses a corpus to confirm or contradict existing descriptive models. Note that the researcher does not validate the pre-formulated descriptive categories. Biber et al. (1998: 10) state that many research questions that are answered with a corpus-based approach should have in fact grown out of other kinds of investigations. This leads to the question of whether the corpus-driven and the corpus-based approach could be linked. The corpus-driven approach is what Tognini Bonelli names 'a qualitative revolution'. She defines it as follows:

> In a *corpus-driven* approach the commitment of the linguist that is to the integrity of the data as a whole, and descriptions aim to be comprehensive with respect to corpus evidence. [...] The theoretical

> statements are fully consistent with, and reflect, the evidence provided by the corpus. (Tognini Bonelli 2001: 84)

As a consequence, the corpus is not only a stock of examples that verify previous assumptions but can be regarded as a blank sheet that serves as a neutral basis for analysis; neutral in so far as a corpus-driven approach can also have implications for the compilation of a corpus. Even if he/she is unaware of the problem, the researcher is in danger of adjusting the selection of data in order to meet existing expectations. Bearing all that in mind, I intend to adopt the corpus-based approach for this study and attempt to answer the research questions by exploring the corpus. For this reason, I have also framed questions rather than hypotheses which have to be proven. Nevertheless, the results gained from a corpus-based approach could also provide the basis for a complementary corpus-driven analysis.

A further corpus-linguistic issue central to this study is the question of *how* the data should be analysed. Corpus analysis can be broadly categorised into qualitative and quantitative analysis. These two types of data analysis form different, but not necessarily incompatible perspectives on corpus data (cf. Biber et al. 1998, Kennedy 1998, McEnery & Wilson 1996). In quantitative research, features are classified, counted and statistically evaluated. The objective of a qualitative analysis is a detailed description of a linguistic feature in context. No attempt is made to assign frequencies to the items identified and, as a result, infrequent phenomena receive the same level of attention as more frequent ones. But what about a combination of both approaches? Biber et al. (1998: 5) claim that a corpus analysis must go beyond simple counts of features as it is "essential to include qualitative, functional interpretations of quantitative patterns". Mukherjee (2004: 112) agrees that once a corpus is used as a linguistic database "it functions as a means to a higher linguistic end. This includes the basic idea that quantitative data in linguistics always require qualitative interpretation". For the purpose of this study, I will therefore choose a multi-method approach that benefits from the advantages of qualitative and quantitative analyses. A quantitative analysis is a valuable precursor to a subsequent qualitative analysis, i.e. before lexical phenomena can then be classified and counted.

The last but crucial question is that of source. A researcher conducting a corpus-linguistic analysis can either draw on existing corpora or can compile a corpus by him/herself. Since the decision for a corpus depends on the research question, I have decided to compile a tailor-made corpus rather than use a subsection of an existing one. The research question is fairly special and concerned with a restricted field of language use. Thus, a self-compiled corpus provides a more accurate and reliable basis for linguistic analysis. Kennedy

(1998) agrees that this should be one of the major reasons for voting in favour of compiling a specialised linguistic corpus.

3.2 Corpus data

3.2.1 Corpus requirements

With the decision to compile a corpus that is tailored to provide appropriate data in order to answer the specific research questions in the restricted field of scientific discourse popularisation, the most significant question to arise immediately is that of corpus requirements. First of all the question of size needs to be discussed. An aspect McEnery and Wilson (1996) even describe as the *pressure* of size. The question of how much data is sufficient to give reliable evidence has to be considered carefully. Hunston (2002: 25/26) approaches the question whether there is an optimal size for a corpus. The LOB and the Brown corpus, for example, contain the once *magic* number of one million words. Kennedy (1998: 73) claims that "there is nothing magic about one million words and no advantage of compiling a corpus of that size if a corpus of 200 000 words will do the job intended". However, there is a tendency towards the creation of mega corpora like the dynamic Bank of English which now contains over 400 million words. Yet, Leech (2001: 10) points out that "to focus merely on the size, however, is naïve". Specialised and highly restricted corpora call for a smaller size, since a special genre and its particularities can be depicted in a better way than many genres or general language. The aim of building a comparable and representative corpus furthermore constrains the choice of material which is to be included. I have therefore decided to aim at a size of approximately half a million words. The amount of data which has to be selected and stored is still manageable and should also guarantee a database that is representative of the two genres and gives sufficient evidence for investigating similarities and deviances, i.e. two subcorpora for medical and popular texts respectively.

The issues of representativeness and authenticity lead to the question of what sources are to be included in the corpus. As far as authenticity is concerned, the written data included in the corpus is a sample of naturally occurring language, that is, texts that were actually published and read by a multitude of readers. Biber (1993) emphasises the importance of defining the limits of the population that is to be studied *before* the actual sampling procedure is dealt with. In order to bring evidence of the different lexical realisations of a scientific topic in the two domains, the corpora have to be in a way representative of discourse in the world of science and in the general public. In this case *representativeness* is not referred to as a sample of language that mirrors its entire diversity but only diversity within this particular setting. In connection to this, Leech (1991: 27) argues that a corpus can be regarded as representative when the evidence gained from it can be "generalized to a larger hypothetical corpus" and Hunston (2002: 30) points out that the real question about representativeness arises when it comes to the interpretation of findings. Although the corpus of the present study is based on topic-related samples of text (i.e. the SARS epidemic), the corpus design aims at being representative of scientific language and general language use respectively. I am well aware of the fact that I cannot make general assumptions about scientific discourse popularisation as a whole by looking at a mere fraction of the situation. However, the texts have been carefully selected and thus the results gained from the subsequent corpus analysis can form a vital part and add to a corpus-based model of popularisation.

Since the research questions demand an analysis of opposing genres (scientific and popular), a comparable sample has to be compiled. The comparability of the corpora implies, despite their genre differences, that the sample criteria agree on some basic features. Therefore, the following criteria have been taken into consideration:

- **medium:** both corpora have to comprise the same medium, in this case written texts, i.e. articles that were published and distributed via up-market print media;
- **topic**: the resulting texts have to be exclusively concerned with a shared subject-matter, i.e. health aspects of the SARS epidemic and not with economic effects of the disease;
- **date:** only texts that were published between February 2003 and October 2003 are to be included;
- **distribution:** the texts should be taken from an equal number of sources that have an international and wide range of distribution, i.e. papers and journals that are edited in the UK and the U.S.A..

3.2.2 Corpus compilation and design

Bearing the corpus requirements in mind, the compilation process requires a decision on particular sources. The main criteria include practical reasons such as free accessibility but also more conceptual ones, such as quality and prestige as well as regional origin of journals, magazines and newspapers. The texts selected for the corpus should not focus on one variety of English alone. For this reason, samples from the UK, U.S.A., Australia, Canada and Asia were collected.

Since scientific journals are the main source of written data in medical discourse, they provide an obvious starting point for the collection of data. The printed versions of all major medical journals are available online as downloadable PDF-files. Thus, a selection could be made according to significant role in health up-date, prestige in a relevant country, international circulation as well as reputation in research. This leads to the selection of the following set of journals from which the texts for the medical subcorpus were taken. Table (3) provides a basic description of the seven journals.

Table 3: *Medical journal description*

Medical Journals	**Description**	**Text types**
WER (Weekly Epidemiological Record)	-Official WHO publication -Published weekly -Important role in health up-date	-Bulletins -Up-dates
The Lancet	-International journal -Published bi-weekly -International circulation	-Research Articles
BMJ (British Medical Journal)	-Journal of the British Medical Association -Published monthly -Prestigious UK journal	-Bulletins -Reports
CMJA (Canadian Medical Association Journal)	-Prestigious Canadian medical journal -Published monthly -In English and French language	-Research Articles
JAMA (Journal of the American Medical Association)	-Prestigious U.S. medical jour-nal -Published monthly	-Research Articles
NEJM (New England Journal of Medicine)	-U.S. medical journal -Published monthly -Reputation in research	-Research Articles
MMWR Morbidity and Mortality Weekly Report	-WHO-affiliated journal -Published weekly -Important role in health up-date	-Reports -Bulletins -Up-dates

Different journals prefer different text types: primarily research-oriented journals feature research articles, whereas journals that aim rather at informing the medical world about current affairs print shorter reports, bulletins and up-dates. A *bulletin* is a statement (e.g. by the WHO) with official status. It usually takes up to one page and contains information about emerging diseases and health threats. *Up-dates* are follow-ups of bulletins to include the latest developments. *Reports* are articles that are not necessarily research-based and provide an overview of a particular topic.

Print media are one of the main sources of information for the broad public and are thus the most important supplier of news on scientific topics. According to MacDonald (2005), they furthermore form a basis for analysing popularisation:

> Readers of more popular news and general magazines have not self-selected to read information about science and are therefore more likely to expect some degree of simplification, vividness, or entertainment with their news. (MacDonald 2005: 278)

Contrary to the medical journals, free access to newspaper and magazine archives is often more restricted. That is why more practical requirements had to be considered for the selection of newspaper articles. The following criteria were established:

- quality up-market papers rather than tabloids,
- daily papers as well as weekly and monthly magazines,
- free online access to print articles,
- a range from news reports to in-depth articles,
- prestige in a relevant country,
- international circulation.

Table (4) provides an overview of the seven papers and magazines that met the criteria and were selected as a source for the popular subcorpus. They were searched for articles related to SARS in the relevant period of time. The text types in the popular press depend primarily on the frequency of publication. Daily newspapers feature news reports and, if a topic is of major interest to the public, leading articles. News reports usually span few paragraphs, whereas a leading article can develop over several pages. Magazines that are published weekly or monthly do not have to operate under immediate pressure of time and can thus provide extensive in-depth reports or even special features on a single topic.

As a pre-requisite for computer-assisted analysis, all texts were downloaded from the respective journal and paper archives as PDF-files which were then converted into plain text files (.txt). The individual text files were then re-edited to delete all types of marginal information that does not belong to the text itself, such as title of the magazine, name of the author as well as location and date (e.g. *London -- 12.06.2003*).

Table 4: *Popular journal description*

Newspapers	Description	Text types
The Guardian & The Observer	Daily quality paper (Sunday - edition is called The Observer) -Published in the UK	-News reports -Leading articles
Time Magazine & Time Asia Mag.	-Monthly magazine -Published in the U.S.A. for the home target market or as an Asian edition for the Asian target market -In-depth reporting on SARS	-In-depth reports -Special issues on SARS
Sydney Morning Herald	-Daily quality paper -Published in Australia	-News reports -Leading articles
Asian Times	-Weekly quality paper -Published in Asia	-News reports -Leading articles -Special issue on SARS
Newsweek	-Weekly magazine -Published in the U.S.A.	-News reports -Leading articles
Washington Post	-Daily quality paper -Published in the U.S.A.	-News reports -Leading articles

The files were grouped according to source and stored in individual folders containing all texts from that source. The texts are thus not stored in a single chunk, but retain their individual structure. In this way, the accessibility of individual texts is guaranteed. Since storage and easy retrieval of data are of central importance for the compilation of a corpus that can also be used by other researchers, iconic codes were assigned to each text. The individual code states the text source and the date of publication (e.g. *Guardian June 16_2*).
After the texts had been stored and edited, a first statistical profile was generated. The two subcorpora comprise the number of articles and running words given in Tables (5) and (6).[6]
The corpora show two prominent features as far as size is concerned which need to be commented on. Both corpora add up to a total corpus size of 498 511

[6] *Words* in this case are the number of overall tokens that were computed by the statistical function of *WordSmith*. The numbers can differ when other programs are used.

words; the popular subcorpus comprises 303 573 words, whereas the medical subcorpus only comprises 194 938 words. The reason for this has to be seen in the different publication routine. The medical journals which feature extensive research articles are only published monthly and the weekly publications include predominately bulletins or up-dates which are smaller in size.

Table 5: *Design of the popular subcorpus*

Newspapers & Magazines	**Articles**	**Words per article**	**Words in total**
The Guardian	162	Ø 633	102 542
Time & Time Asia Magazine	42	Ø 1 205	50 628
Sydney Morning Herald	31	Ø 520	16123
Asian Times	11	Ø 1 898	20 879
Newsweek	8	Ø 858	6 860
Washington Post	102	Ø 1 043	106 418
in total	**356**	**Ø 853**	**303 573**

Table 6: *Design of the medical subcorpus*

Medical Journals	**Articles**	**Words per article**	**Words in total**
WER (Weekly Epidemiological Record)	19	Ø 800	15 188
The Lancet	46	Ø 1 362	62 651
BMJ (Journal of the British Medical Association)	20	Ø 771	15 426
CMJA (Canadian Medical Association Journal)	12	Ø 1 631	19 567
JAMA (Journal of the American Medical Association)	8	Ø 2 819	22 552
NEJM (New England Journal of Medicine)	14	Ø 2 282	31 954
MMWR (Morbidity and Mortality Weekly Report)	23	Ø 660	15 188
in total	**142**	**Ø 1 273**	**194 938**

As I am particularly interested in the popular representation, the medical subcorpus serves as a reference corpus and has hence a legitimatisation for being smaller. Regarding the analysis, this difference in size has to be kept in mind and normalised frequency is employed as the number of occurrences will be given in percentages.

3.3 Methodology II: Tools and Analysis

3.3.1 Computer-assisted analysis

Nowadays the term *corpus* nearly always implies the idea of machine-readability. As outlined previously, this was not always the case as in the past the word *corpus* was also used in reference to non-computerised texts. Machine-readable corpora have many advantages over written or spoken formats as they can be searched and manipulated at speed when analysed electronically. The combination of computers and corpora set the ground for a new view on language. Sinclair (1991: 100) remarks that "language looks rather different when you look at it at once", but for managing large amounts of data, computer-assisted analysis is a crucial factor:

> Strictly speaking, a corpus by itself can do nothing at all, being nothing than a store of language. Corpus-access software, however, can re-arrange that store so that observations of various kinds are made. (Hunston 2002: 3)

The corpus compiled for this study is regarded as a digital database of half a million words which can be explored automatically. The lexical information can be extracted from the corpus with the help of *WordSmith* Version 4 developed by Mike Scott (2004). According to Scott, *WordSmith* is an integrated suite of tools for looking at how words behave in texts. The main functions used in this analysis include *WordList*, *StopList*, *Keywords*, *Concord*, *Collocations* and *Clusters:*

WordList: The *WordList* tool computes a list of all the words in a text, set out in alphabetical or frequency order. The word lists presented in this study are all given in frequency order. The list can be used for identifying the words used in a corpus but even more so to compare the frequency of words in multiple corpora.

StopList: The *StopList* is a list of common grammatical/ function words which are to be excluded from the results displayed by a word list. This helps to pinpoint the lexical/ content words only. To use this function, a file has to be prepared specifying all the words that have to be ignored.

Keywords: With the *Keyword* function, *WordSmith* provides a means for identifying statistically differing words in (two) collections of texts. A word is identified as *key* by making a direct comparison between one wordlist (from the target corpus) and another reference worldlist. The keywords are either unusually frequent in the target corpus in comparison with the reference corpus (i.e. *positively* key) or are unusually infrequent in the target corpus (i.e. *negatively* key).

Concord: The program generates concordance lines for a specified search word or any word-list item. It displays the immediate context of a word (five words to the right and to the left) and gives access to information about collocates or clusters of the particular word. The concordance lines are presented in the KWIC (keyword in context) format. Hunston (2002: 9) points out that the analysis of concordance lines can display regularities that otherwise remain unobserved when the word is met in its textual context.

Collocation & Clusters: Collocates are the words which occur in the neighbour-hood of a search item. The function operates statistically in order to compute the most likely co-occurring words. In this way, a collocational profile can be created. A cluster is a group of words which follow each other sequentially in a corpus. According to Firth (1957) a closer look at the collocates helps to find out more about the company a word keeps and this tells us more about its meaning and its usage.

3.3.2 Analytical steps

Methodologically, the corpus will not be searched systematically, but rather explored by using a multi-method approach of combining a quantitative with a qualitative/ functional analysis. To answer the research questions sketched out in 2.3, at least a framework of possible analytical steps needs to be developed before the corpus is browsed. The analysis is thus divided into individual steps that will be outlined in the following.

QUANTITATIVE ANALYSIS

Basic frequency: All the words in the corpora will be arranged in order of their frequency. This is most interesting, because frequency can give first insights into the distribution of words or items in the particular genres. This is done for each corpus individually.

Refined frequency: Since the objective of this analysis is to bring light to relevant content words only, the previous results can be further refined by applying a filter technique provided by the *StopList* function. The lexical words in a corpus can reflect the subject matter. I compiled this list by extracting the 50 most frequent words from the FLOB-corpus (Freiburg Lancaster-Oslo/Bergen).[7] I intend to compute the list from the existing FLOB-corpus rather than compiling one myself, as it provides an unbiased and neutral account of the most frequent lexical items in a balanced corpus of written samples. The FLOB is the updated equivalent of the LOB corpus and comprises about 1 million words of written British English. Additionally, the stopwords extracted from the FLOB-corpus are highly functional and grammatical items (e.g. *the*, *of*, *and*, *to*, *a*, *in*, *that*, *is*, *was*, *for*).

Comparison of the two refined lists: The refined lists are particularly useful for comparing content words of the two corpora to show a common stock of words or deviances. Therefore, the two lists are matched and compared in terms of their 100 most frequent words and in terms of their low frequency keywords. Words in this case have to be seen as tokens, as it is of interest to include the individual realisations of types in the analysis. The differences and similarities can then be studied in more detail.

QUALITATIVE ANALYSIS

The preceding results of the frequency analysis can now be investigated from a qualitative point of view by examining the context of selected phenomena. The examples are chosen according to their representativeness in terms of frequency which has been established in the quantitative analysis.

Concordance lines: The words or lexical items are interpreted with the help of their particular context in order to verify potential similarities and/or differences in genre-specific usage. Lexical patterns and structures in which the word occurs will be identified and characterised. Since the total number of concordance lines produced by *WordSmith* exceeds the number of lines that can be displayed by far (e.g. 1740 concordance lines for *patients*), the tables will present unsorted extracts in KWIC format.

Collocational profiling: The interpretation of individual usage will be supported and complemented by generating and looking at selected collocational profiles. The immediate context of the respective words can give evidence of particular tendencies as far as co-occurrence and semantic prosody is concerned. Furthermore, the function of an item can be presumed from its collocates or clusters. This leads to an investigation of the similar or deviant usage of words in both corpora.

---

[7] The FLOB corpus was compiled by Christian Mair and his team at the University of Freiburg in the early 1990s.

FUNCTION AND RELATION
In a final step, the functions that the words that have been analysed quantitatively and qualitatively fulfil in scientific and popular discourse will be identified ("function") and compared in the two subcorpora ("relation"). The findings will then be explained in terms of relevant variables. This leads to the establishment of a model that pictures similarities and deviances between the two genres.

3.4.Pilot study: exploring the corpus

3.4.1 Why conducting a pilot study?

The corpus and the methodology developed for analysing the data were developed on the basis of theoretical preliminaries. The corpus-based approach calls for exploring the corpus without a sort of guidance that could have been provided, for example, by predefined queries. Before going *in medias res* with the analysis of the large corpus, it makes sense to put it to an initial test. It helps to answer questions and eliminate doubts, such as the following: Will the data provide sufficient evidence for making assumptions about relations between the two genres? Are the analytical steps useful for getting significant results?

In conducting a pilot study and analysing a smaller corpus, the researcher has the opportunity to approach some hypotheses and techniques before the large-scale analysis is conducted. Hunston (2002: 25) argues that the sheer quantity of information alone can be overwhelming for an observer and that extracting information from concordance lines that go into the thousands is hardly manageable. She also suggests as one solution to try a smaller corpus first.

The pilot corpus used for this pre-study is a sample taken from the full corpus compiled for the subsequent large-scale analysis. The pilot corpus is a random sample of 42 files was taken from the selected texts. It is structured in two parts: a scientific subcorpus compiling medical journal articles and a popular subcorpus consisting of general newspaper and magazine articles. It comprises a selection of texts in which each publication is represented by three articles. Table (7) provides an overview of the pilot-corpus design. Each subcorpus comprises 21 texts taken from seven different sources. The divergent number of words, about 25 500 in the popular corpus against about 43 500 in the medical corpus, is a direct result of the genre characteristics of the different text types. As outlined previously, scientific research articles span several pages, whereas newspaper articles are usually much shorter.

Table 7: *Design of the pilot corpus*

Corpora	Text Sources	Articles	Words
Subcorpus I: Popular Corpus	Newspapers & Magazines : ▪ Asia Times ▪ Guardian ▪ Newsweek ▪ Sydney Morning Herald ▪ TIME Magazine Asia ▪ TIME Magazine International ▪ Washington Post	21	25 459
Subcorpus II: Medical Corpus	Medical Journals: ▪ WER (Weekly Epidemiological Record) ▪ BMJ (British Medical Journal) ▪ The Lancet ▪ JAMA (Journal of the American Medical Association) ▪ CMAJ (Canadian Medical Association Journal) ▪ MMWR (Morbidity and Mortality Weekly Report) ▪ NEJM (New England Journal of Medicine)	21	43 424

3.4.2 First steps, first results

The first step of the analysis is to calculate a basic ranking that gives the 30 most frequent words in each corpus separately. The results are given in Table (8).

Table 8: *Basic frequency ranking*

N	Popular SC	Medical SC	N	Popular SC	Medical SC
1	THE	THE	**16**	ARE	WHO
2	OF	OF	**17**	VIRUS	CASES
3	TO	AND	**18**	WAS	IS
4	A	IN	**19**	AS	AT
5	IN	TO	**20**	BEEN	DAYS
6	AND	PATIENTS	**21**	WHO	THAT
7	IS	A	**22**	AT	HAVE
8	THAT	SARS	**23**	BY	BY
9	SARS	WITH	**24**	ON	HOSPITAL
10	IT	WAS	**25**	HONG	PATIENT
11	HAVE	FOR	**26**	KONG	CASE
12	BE	WERE	**27**	BUT	AS
13	FOR	ON	**28**	DISEASE	RESPIRATORY
14	FROM	OR	**29**	THIS	THIS
15	WITH	HAD	**30**	NOT	FROM

Table 9: *Detailed frequency list after applying StopList*

N	Popular SC	Freq. %	Medical SC	Freq. %	N	Popular SC	Freq. %	Medical SC	Freq. %
1	SARS	1.22	PATIENTS	1.60	16	BECAUSE	0.19	FEVER	0.22
2	VIRUS	0.61	SARS	1.26	17	HOW	0.19	ADMISSION	0.21
3	HONG	0.47	CASES	0.50	18	VACCINE	0.19	REPORTED	0.20
4	KONG	0.47	DAYS	0.47	19	SCIENTISTS	0.18	ILLNESS	0.20
5	DISEASE	0.44	HOSPITAL	0.42	20	WORLD	0.18	MARCH	0.20
6	PATIENTS	0.37	PATIENT	0.42	21	OUTBREAK	0.18	DURING	0.19
7	PEOPLE	0.34	CASE	0.40	22	VIRUSES	0.16	PROBABLE	0.19
8	HEALTH	0.30	RESPIRATORY	0.39	23	HOSPITAL	0.16	AMONG	0.18
9	CASES	0.30	CLINICAL	0.31	24	SINGAPORE	0.16	SEVERE	0.18
10	INFECTED	0.24	SYMPTOMS	0.28	25	SYMPTOMS	0.16	INFECTION	0.17
11	CHINA	0.22	CARE	0.27	26	AGAINST	0.15	SUSPECTED	0.17
12	SPREAD	0.22	HEALTH	0.25	27	OFFICIALS	0.15	HONG	0.17
13	CORONAVIRUS	0.21	CHEST	0.25	28	RESPIRATORY	0.15	KONG	0.17
14	COV	0.20	DISEASE	0.24	29	LAST	0.15	CORONAVIRUS	0.16
15	EPIDEMIC	0.20	CONTACT	0.24	30	KNOWN	0.14	TRANSMISSION	0.16

Since the objective of this analysis is to highlight the relevant key words only, the previous results can be further refined by applying a filter technique provided by the *StopList function*. With the help of this technique, the 50 most frequent (function) words from the FLOB-corpus were extracted and stored as stop words. Table (9) gives the results after activating the *StopList-function* and re-computing the previous word list. Since only the most frequent content words are displayed, the stoplist is certainly a step worth taking to facilitate further analysis. The lists now present a valid snapshot of the lexical inventory of both genres, marked by their reflection of the same restricted topic. Similarities and differences of the two discourse forms are identified in the following by comparing the resulting 30 most frequent content words of the two subcorpora. The most obvious procedure to detect similarities and differences is to match the two lists and detect the words that are found in both corpora and those that are unique to only one subcorpus. Table (10) groups the words into those that appear in both registers (11), and those that are only used in the popular discourse (19), or in its medical counterpart (19).

Table 10: *Word similarities and differences*

Shared words	**Words unique to the popular corpus**	**Words unique to the medical corpus**
11	19	19
SARS, PATIENTS, CASES, DISEASE, HONG, KONG, HEALTH, HOSPITAL, CORONAVIRUS, RESPIRATORY, SYMPTOMS	VIRUS, PEOPLE, INFECTED, CHINA, SPREAD, COV, EPIDEMIC, BECAUSE, HOW, VACCINE, SCIENTISTS, WORLD, OUTBREAK, VIRUSES, SINGAPORE, AGAINST, OFFICIALS, LAST, KNOWN	DAYS, PATIENT, CASE, CLINICAL, CARE, CHEST, CONTACT, FEVER, ADMISSION, REPORTED, ILLNESS, MARCH, DURING, PROBABLE, AMONG, SEVERE, INFECTION, SUSPECTED, ASSOCIATED

Contrary to the dominant view of science popularisation discussed in section 2, a large number of words seems to be shared by both discourse communities. However, this simplified categorisation can be further refined by discussing the distributional profiles of the respective types, both quantitatively and qualitatively. If we refine the quantitative analysis by studying their frequency profile more closely, the words can be grouped into the following categories:

I The word appears in only one of the two corpora in high frequency:

- particular to the popular subcorpus: e.g. outbreak, officials

- particular to the medical subcorpus: e.g. care, contact

II The word appears in both corpora with a frequency that is:
- almost equal: e.g. SARS, health
- higher in the popular than in the medical corpus: e.g. virus, Hong Kong
- lower in the popular than in the medical corpus: e.g. respiratory, symptoms

III A similar concept or token appears:
- same type, different token: e.g. patients in the popular corpus and patient in its medical counterpart
- related concept, i.e. a hyperonym: e.g. virus in the popular corpus and coronavirus in the medical corpus

When we take tokens and sense-related items into account, an additional part of the potential ties between the two registers becomes apparent. Both subcorpora include similar concepts by using different realisations, for example for referring to affected persons by using patient, patients, case, cases, or people.

Table 11: *Shared concepts I*

	Popular SC	Freq. N / %	Medical SC	Freq. N / %
PATIENT	-		+	6 / 0.42%
PATIENTS	+	6 / 0.21%	+	1 / 1.60%
CASE	-		+	7 / 0.40%
CASES	+	9 / 0.16%	+	3 / 0.50%
PEOPLE	+	7 / 0.16%	-	

On the one hand, Table (11) clearly shows that similar notions are expressed in both corpora, but, on the other hand, differences in the realisation become obvious: In the popular corpus only plural forms are used and, according to the genre-specific expectations, the general term people is not used in the medical corpus. These aspects will be explained in the qualitative analysis section in more detail.

First of all, a further example of shared concepts that does not act in agreement with the genre-specific expectations will be analysed. For the underlying subject matter of SARS, the concept of a causative agent, i.e. a bacteria or a virus, is of importance. Since this is first and foremost a medical fact, one would expect a very frequent representation of the concept. Table (12) shows that the notion of causative agent can be realised by using virus, viruses, coronavirus, or SARS-COV. Contrary to expectations, the words virus and viruses are not found within the most frequent items in the medical corpus. The specific virus causing SARS, the coronavirus, is present in both discourse forms, but it occurs significantly more frequently in the popular corpus than in its medical counterpart.

Table 12: *Shared concepts II*

	Popular SC	Freq. N / %	Medical SC	Freq. N / %
VIRUS	+	2 / 0.61%	-	
VIRUSES	+	22 / 0.16%	-	
CORONAVIRUS	+	13 / 0.21%	+	29 / 0.16%
SARS-COV	+	14 / 0.20%	-	

Furthermore, the abbreviation SARS-COV appears in the popular corpus only, although LSP linguists regard abbreviations and acronyms as a scientific-genre marker (cf. Beier 1980, Fluck 1991, Sager et. al 1980).

As far as the quantitative aspect of the interrelation of both registers is concerned, the frequency analysis has shown that particular sets of characteristic lexical items exist: one set that is subject-matter specific but unique to each genre, and another that is also subject-matter-specific but shared by both discourse types. They also draw on similar concepts or referents, though the corpora do not necessarily display identical lexical realisations.

QUALITATIVE ANALYSIS

The preceding results of the frequency analysis are now investigated from a qualitative point of view by computing and examining concordance lines for selected examples. The examples have been chosen according to their representativeness in the three major categories established before. The relevant word or concept is thus interpreted with the help of its particular context verifying potential similarities and/or differences in the genre-specific usage. The following contextual analysis aims at providing information on how and why particular words are used in the individual registers.

Category I: Words that appear in only one of the subcorpora in high frequency

The frequency analysis has shown that certain lexical words are unique to the popular subcorpus. One prominent concept found in a consistently high frequency is that of an epidemic or an outbreak that affects the world (*epidemic* = 0.20%; *outbreak* = 0.18%; *world* 0.18%). By looking at the word *outbreak* in context, further genre-specific aspects become apparent. As Figure (8) illustrates, *outbreak* co-occurs with words and concepts that have explicit negative connotations and thus show a negative semantic prosody (cf. Louw 1993, Sinclair 1991, Partington 1998).

N	Concordance		
1	as a travelling killer. The	*outbreak*	and rapid spread of the
2	travel warnings: owing to the	*outbreak*	the U.S. State Department
3	soon after the first SARS	*outbreak.*	I even had to cancel
4	to be the same illness in an	*outbreak*	that began last November
5	the cause of the mysterious	*outbreak.*	WHO officials also said
6	look at its own earlier	*outbreak,*	which killed five people
7	of the patients in the latest	*outbreak*	seem to be recovering.
8	of a written summary of its	*outbreak*	was an unprecedented
9	alert, fearful of a worldwide	*outbreak*	of a highly contagious
10	the epicentre of the SARS	*outbreak*	outside of the mainland

Figure 8: *Concordance sample for* outbreak *in the popular corpus*

SARS is pictured as a *traveling killer* (line 1), as an *outbreak which killed people* (line 6), and which is *mysterious* (line 5). Consequently, the audience learns about *travel warnings* and a *state alert* (lines 2, 9). A cross-check for the adjective *fearful* (line 9) with the medical subcorpus confirmed that there is not a single occurrence of either *fearful*, or the token *fear* in the scientific counterpart. Not only are the concepts of epidemic and outbreak prominent features of the popular representation of the topic in the general press, but they are furthermore combined with emotionally appealing formulations. This is opposed to the absence of such formulations and concepts in the medical journals. Semantics prosody is thus a prominent pattern that should be investigated in detail in the large-scale analysis

As Dirckx (1988) explains, medical language is characterised by neutrality, factuality, and accuracy. This means that scientific terms are predominantly used because they are free of connotations, they are unambiguous and provide restricted semantic reference to the medical professional. The concordance line of the word *contact*, which occurs frequently (0.24%) only in the medical corpus, illustrates that general concepts are modified accordingly in order to adopt the characteristic attributes of neutrality, factuality and accuracy.

The first three examples in Figure (9) show that the general concept of *contact* is specified by using the premodifying adjectives *close* (line 1), *face-to-face* (line 2) and *direct* (line 3). The notion of *face-to-face* is even further refined by adding the concrete measure *within 1 m* (line 2). This importance of defining the concept of contact is indicated in line (4) by the context words *type*, *duration* and *proximity*. Figure (9) also underlines the significance of interpreting words on a qualitative basis: the concordance lines indicate that *contact* fulfils a dual purpose, since it is also used as a premodifier for the concept of precautions (*contact and droplet precautions*; lines 6 to 9).

N	Concordance		
1	SARS cases or to have close	*contact*	with a suspect or probable
2	or had face-to face (within 1m)	*contact*	with a person with SARS
3	person with SARS or had direct	*contact*	with respiratory secretion
4	type, duration and proximity of	*contact*	and infection control
5	120 (93,8%) had documented	*contact*	with a SARS case or with
6	in his differential diagnosis.	*contact*	and droplet precautions
7	using airborne, droplet and	*contact*	precautions. No further
8	who did not immediately use	*contact*	and droplet precautions.
9	department, where airborne,	*contact*	and droplet precautions
10	number of people who were in	*contact*	with case B and his wife

Figure 9: *Concordance sample for* contact *in the medical corpus*

A qualitative analysis can provide further insights into function and actual usage and in this way complement the frequency analysis.

Category II: Words that appear in both corpora

Despite all the apparent differences between both registers, the frequency analysis has proven that both subcorpora share over one third of their most frequent lexical items, implying that a common core exists. Nevertheless, the previous example has suggested that this similarity might not be reliable as far as function is concerned. This is now put to the test by comparing their concordance lines and lexical patterns for the word *health*. It is found with similar high frequency in the popular (0.30%) and in the medical subcorpus (0.25%).

First, in Figure (10), the concordances for *health* in the popular subcorpus are given. The remarkable salient feature that becomes obvious by looking at the word *health* in context is that the word is consistently used as a constituent of a multi-word expression and not as a distinct noun. According to corpus- and usage-based dictionaries, such as the *Collins Cobuild Dictionary of the English Language* (1991) or the *Macmillan English Dictionary for Advanced Learners* (2002), the word *health* is commonly refers to a state or a condition of the body with the typical collocations *being in good* or *poor health*. Surprisingly, this concept seems to be of minor significance for the popular representation of the SARS discourse in the general press. The *Macmillan Dictionary* gives, in addition, a separate entry for *health care* referring to the service responsible for people's health. This meaning appears to correspond to the realisation of health in the popular subcorpus.

N	Concordance		
1	no longer new. Scientists,	*health*	-care professionals and
2	faster than the ability of the	*health*	care system to find and is
3	that the mainland Chinese	*health*	authorities, by failing to
4	data are not correlational,	*health*	authorities may be able to
5	outbreak of SARS did occur,	*health*	-care workers would need
6	it is another for a rural	*health*	clinic in the Chinese or
7	who visited an Environmental	*health*	institute (EHI) lab, where
8	is good news for the public-	*health*	authorities, because it
9	and prevention. Which public-	*health*	measures will help most to
10	will become more difficult for	health	officials to monitor l

Figure 10: *Concordance sample for* health *in the popular subcorpus*

Figure (10) shows that the clusters incorporating health most frequently relate to a specific group or body (line 1: *health-care professionals*, line 5: *health-care workers*, line 10: *health officials*, lines 3, 4, 5: *health care authorities*). Since this realisation of *health* is marked as a particular and distinct feature, it is of interest to compare the results of the popular representation with its medical counterpart.

N	Concordance		
1	using criteria from the World	*Health*	Organisation and transfer
2	visitors (14.1%) or other	*health*	-care worker (5.5%). Of the
3	case definition issued by	*Health*	Canada. 7 suspected cases
4	symptom onset, to a World	*Health*	Organization (WHO) report
5	were reported to local public	*health*	authorities by clinicians
6	cases or their proxies by	*health*	-care professionals. Data
7	remaining 8 cases, including	*health*	-care workers and patients
8	that was spreading among	*health*	-care workers in Hanoi,
9	involved in the local public	*health*	unit suspected the family
10	cases was reported to public	*health*	officials, 17, 18, which

Figure 11: *Concordance sample for* health *in the medical subcorpus*

Figure (11) shows identical patterns for form and usage. *Health* is also used as a defining constituent in multi-word expressions referring in the same way to *health (care) workers*, *professionals* and *officials*. Furthermore, relevant bodies, such as the *World Health Organization* (lines 1, 4) and *Health Canada* (3) are specified. The features are shared by both genres and clearly deviate from the usage norm listed in the corpus-based dictionaries. This leads to the conclusion that they are subject-matter and not genre-specific. The preceding example has revealed that both discourse forms share not only lexical forms in a similar frequency but that they can also correspond in topic-specific use of those forms. But what about lexical forms that are also shared but which differ in frequency, such as *respiratory*? It is the eighth most frequent content word in the medical subcorpus (0.39%) and holds the 28th rank in the popular subcorpus (0.13%). In other words, *respiratory* takes a very important role in the medical representation. It is vital to find out whether this significance is also reflected in the concordance lines. First, in Figure (12), a closer look at the popular subcorpus is taken.

N	Concordance		
1	his mother had died of a	*respiratory*	illness, and his back
2	be behind a dangerous new	*respiratory*	infection spreading among
3	common colds and sometimes	*respiratory*	infections in premature
4	like colds an upper-	*respiratory*	tract infections, and
5	known as severe acute	*respiratory*	syndrome (SARS), is
6	epidemic of a severe	*respiratory*	ailment known as SARS
7	SARS stands for severe acute	*respiratory*	syndrome. It's a new and
8	including mumps, measles and	*respiratory*	illnesses. Coronavirus
9	stricken by severe acute	*respiratory*	syndrome are believed to
10	rates of most other	*respiratory*	illnesses. "The number of

Figure 12: *Concordance sample for* respiratory *in the popular subcorpus*

Figure (12) illustrates that *respiratory* either forms part of the expression *severe acute respiratory syndrome* (lines 5, 7, 9) or it denotes a particular type of *illness*, *infection*, or *ailment* (lines 1-4, 8, 10). On the one hand, it has a determining function, but on the other hand, it is also modified by preceding elements like, for example, the preposition *upper* or the adjectives *severe* and *acute*. These findings imply that *respiratory* in the popular corpus is used in accordance to its scientific origin, meaning that it is unambiguous and well-defined. However, the context words *other* (line 10) and *sometimes* (line 3) indicate a certain vagueness that is further supported by the use of the indefinite article *a* (line 1) and the use of the plurals *illnesses* (lines 1, 8, 10) and *infections* (lines 3, 4). The concordances also show that *respiratory* co-occurs with words that show a negative semantics prosody, carry negative connotations and assign hazardous attributes to the concept: people are *stricken* (line 10) and *die* (line 1), as it is *new* and *dangerous* (line 2). If those features are also subject- and not genre-specific, they should be found in the concordance of *respiratory* in the medical subcorpus as well.

Figure (13), however, shows that *respiratory* is not only preceded by specifications like *all* (lines 2, 4, 5) and the definite article *the* (line 8), but it is also used in different contexts that do not allow for incorporating negative connotations. It is likewise used as a modifying adjective, but not for the concepts of illness or infection. The realisation of *respiratory* in the scientific context is frequently used for determining the type of *specimen* (lines 2-5), that is, the kind of sample that was taken and analysed, or *syncytial virus* (line 6, 9).

N	Concordance		
1	were prescribed due to severe	*respiratory*	failure (Pa=2 < 50 m
2	Pneumocystiy jiroveci on all	*respiratory*	specimens was negative
3	were negative for blood,	*respiratory*	and urine specimens
4	against Legionella spp. on all	*respiratory*	specimens, and tested
5	DNA was extracted from all	*respiratory*	specimens. Polymerase
6	1/2/3, adenovirus and	*respiratory*	syncytial virus was
7	patients with more severe	*respiratory*	impairment consisted
8	the nursing staff and the	*respiratory*	therapists monitored
9	approved for the treatmemt of	*respiratory*	syncytial virus and
10	course was complicated by	*respiratory*	insuffiency, there was

Figure 13: *Concordance sample for respiratory in the medical subcorpus*

In addition, it refers to a respiratory condition, for example, *insufficiency* or *failure* (lines 1, 10). The comparison of the concordance lines has illustrated that, despite their sameness in form and basic modifying function, fundamental differences as far as the correlating concepts are concerned do in fact exist. This helps to explain the frequency difference since *respiratory* is used in a different conceptual frame in both corpora: In the medical discourse of SARS, the function of specifying the type of specimen, virus, or respiratory condition is of major significance, which is also reflected in the concordance lines.

Category III: Concepts that are realised by using a similar concept or token
The discussion of the last category relates back to the observations already made in 3.1. The quantitative analysis has revealed a stock of words that is shared by both discourse communities and that are realised as same types or even as corresponding tokens. Both subcorpora include, for example, the idea of persons affected by SARS, which is reflected in the lexical concept of *patients* and *people*. The frequency analysis has highlighted that *patient* in its singular form is only found in the neighbourhood of the most significant words of the medical subcorpus (0.42%), whereas its plural form *patients* is highly frequent in the medical (1.60%) subcorpus and is also the 6th most frequent content word in the popular subcorpus (0.21%). This interrelation can be further defined by a complementary analysis of concordance lines for the word *people*, which is the 7th most frequent word in the popular subcorpus (0.16%). First, the concordance for the plural *patients* is given in Figure (14) for the popular part.

The concordance lines indicate first of all that *patients* refers to an indefinite group of *people* that are *sick* (lines 6, 10) and *suffer* from *SARS* (lines 1, 5, 8). The co-occurring words emphasise their passive role and highlight certain actions that are done to and with the patients. They are *treated* (lines 2, 5), isolated (lines 4, 10) and *cared for* (line 9) so that they can eventually *recover* (line 7).

N	Concordance		
1	stereoid drugs in some SARS	*patients*	, stereoids typically do
2	precautions when treating	*patients*	With what was an unknown
3	and throat swabs from	*patients*	Arriving in their surgeries
4	At Imperial College, London.	*Patients*	Have been treated with
5	Workers who'd treated SARS	*patients*	Stayed home in quarantine
6	and the isolation of sick	*patients*	Can slow the spread of
7	To worry about. Though most	*patients*	Recover, SARS can kill
8	an avian coronavirus in SARS	*patients*	. Coronavirus circulates
9	weren't all caring for	*patients*	. So how did they contract
10	For. Besides isolating sick	*patients*	, a number of countries

Figure 14: *Concordance sample for* patients *in the popular subcorpus*

However, no specific words or figures are given and general statements using vague expressions are made, ***drugs in some SARS-patients*** (line 1), ***most patients recover*** (line 7). If we compare this with the concordances for patients in the medical subcorpus, the significance of this difference becomes obvious.

N	Concordance		
1	to have suspect SARS. Of the 4	*patients*	(cases 1,3,8 and 9) were
2	4 (SD 3) days. For the 10	*patients*	with multiple contacts
3	with multiple contacts with	*patients*	with SARS, the intervals
4	3.5 (SD 3) days. Among our	*patients*	, fever, headache,
5	illustrated in Fig. 1. All	*patients*	had a temperature that
6	headache was presented in 7	*patients*	(50%), and 6 (42%)
7	nausea was present in 3	*patients*	; none had diarrhea.
8	was experienced by all	*patients*	during their first week
9	and vomiting developed in 7	*patients*	2-3 days after admission
10	did develop on day 4 in 3	*patients*	. 6 patients required

Figure 15: *Concordance sample for* patients *in the medical corpus*

The most salient and consistent feature in Figure (15) is the accuracy by which *patients* is specified. *Patients* are not an immeasurable group of affected people, but the definite number or a denomination (e.g. *all* in line 5 and 8, *our* in line 4) is always given. Additionally, patients that share an important feature are grouped together. The nature of this characteristic is also defined by the context words, since *patients* correlates frequently with specific symptoms, such as *fever* or *temperature* (lines 4, 5), *headache* (lines 4, 6), *anorexia* or *nausea*, and *vomiting* (lines 7,9). The findings indicate that the concept of *patients* is highly significant for both registers, but that it is used in a different way (general vs. specific) depending on the contextual frame (action vs. symptoms). In sum, this interpretation helps to explain why the singular form patient is only found as frequently in the medical corpus.

N	Concordance		
1	with the highest values of LD, 1	*patient*	(case 3) had minimal
2	also an increase in LD. In the 1	*patient*	with elevated CK and
3	was noted in 2 patients, and 1	*patient*	showed QT prolongation
4	prolongation (0.49 seconds). One	*patient*	developed atrial fibril
5	(179 beats per minute). This	*patient*	responded to medical
6	at least twice a week. One	*patient*	(case 19) has returned
7	cause it before the illness. One	*patient*	continued to have mild
8	on admission in only 1	*patient*	, it became elavated
9	however, in our cohort only 1	*patient*	patient had type 2 diab.
10	contact with a known SARS	*patient*	. Probable SARS requires

Figure 16: *Concordance sample for* patient *in the medical corpus*

In Figure (16), the concept of a group of patients that share a certain feature is now reduced to a minimum, that is, to one single patient. Consequently, *patient* in its singular form is used to refer to his or her individual performance and specific outcome. The particularity of this single patient is also reflected by introducing the concept of *case* plus *number*, e.g. the *patient* equals *case 3* (line 1), the *patient* equals *case 10* (line 6). The consistency of the examples underlines the importance to distinguish between individuals. Bearing this in mind, the findings have shown that this is only relevant to the medical discourse but not to the general public. The popular subcorpus confirms a tendency to generalise, because *people* ranks among the top ten frequent content words.

The direct comparison of the concordances of Figure (15) with those of Figure (17) shows that *patients* and *people* share similar features and concerns. Both are subject to *diagnosis*, *quarantine* and *recovery* (lines 1, 5, 10), but they can also actively *spread* (line 3) the disease when they *sneeze* or *cough* (line 9). A consistent feature of the popular discourse are vague quantifiers, such as *some* and *many* (lines 2-4, 10), which are not used in the medical counterpart. Numbers are only given, if they convey an impressive amount, such as *1,700 people* who have been quarantined (line 1), or *20,000* who have been killed by the flu (line 8).

N	Concordance		
1	Singapore, there, nearly 1700	*people*	have been quarantined
2	drawn the conclusion that some	*people*	don't seem to spread it
3	seem to spread it to many	*people*	while others spread it to
4	salvia, which is why so many	*people*	in Asian cities are now
5	appear to be most at risk.	*People*	diagnosed with SARS
6	much about contracting SARS.	*People*	who live in hard-hit areas
7	keep their distance from	*people*	known to be infected
8	the flu, which kills 20,000	*people*	a year in the U.S. alone.
9	direct, close contact when	*people*	sneeze or cough, they
10	breathing, some get diarrhoea.	*People*	start to recover after

Figure 17: *Concordance sample for* people *in the popular subcorpus*

Hence, the concordance has proven that the concept of affected persons can be realised by using either *patients* or *people*, because here they are nearly synonymous. The analysis of this category further indicates that despite their relation in form, the individual use of the similar concepts is genre-specific and is determined by the expectation and needs of the target audience. Medical professionals rely on accurate results and individual performance, whereas the notion of generality unites the public at large. Results are only then presented if they are relevant to the average reader, for example, by including outstandingly high numbers of newly infected people.

3.4.3 Implications for the present study

As far as the scientific and popular coverage of the SARS epidemic is concerned, the pilot study has revealed that there is a point of intersection between the two distinct levels of discourse. The world of scientists and the general public cannot be regarded as two opposing spheres but are clearly intertwined. The quantitative study has shown that they in fact share a large amount of lexical concepts on a high-frequency basis. However, the results of the qualitative analysis further indicate that there is variation within this common stock of words, since function and usage of the individual concepts can differ. The sheer presence of a particular vocabulary item in both subcorpora does not bear evidence for similar usage, nor does the absence of a word indicate the absence of the corresponding concept. The discussion of the concordance lines reveals that what is communicated is not necessarily identical to the respective structure, e.g. the type patients is similar in form but is embedded in different contexts with different meaning and significance to each audience. The analysis further underlines the necessity to investigate in a broader context, that is, the subject matter or topic. The example of the identical form and function of ***health*** demonstrates that this external factor accounts for patterns that clearly derive from the (usage) norm in general language. Apart from subject-specific features, genre-specific realisations seem to play a key role in both discourse types. Deviations in the individual realisation of shared lexical concepts apparently result from essential and unshiftable characteristics of the respective genre. Special languages have to be precise, unambiguous, and free from emotional appeals or semantic prosody. The presentation of SARS in the popular press, on the other hand, appears to be not only bound to the transfer of information and facts, but it also depends on genre particularities, such as 'bad news is good news', meaning that negative news sell. This could explain the high amount of words with negative semantics prosody and associations, e.g. *fearful*, *killer virus*, or *worldwide outbreak* in the popular subcorpus. The next step in refining these preliminary results is, of course, to transfer them to the large corpus and put them to the test. The pilot study provides a valuable basis for the following large-scale analysis as it points the way in which it should be directed. Although studies of lexical phenomena require large corpora in order to make conclusive statements, preceding small-scale studies are a step worth taking in order to put the compiled corpus and the basic methodology to a first test.

4 Quantitative analysis

4.1 Defining the word

The obvious point of departure in the description of lexical processes is the definition of the unit if analysis itself, that is, the word. Although it is intuitively known what is meant by a word, the formal description of wordhood is not as simple as it appears, since the concept is neither straightforward nor universal. So far, I have used *word* or *lexical item* without specification of the definition. When the terms are examined more carefully, it becomes evident that different approaches have to be taken into account. Some approaches can be fruitfully combined while others are mutually exclusive. I would nevertheless argue that the multiple meanings of the term *word* reflect its multidimensional nature rather than its vagueness. Researchers working in different traditions will employ different terms, so the very dimension of a word which is referred to depends on the linguistic perspective and the individual context.

Despite all differences, linguists agree that the term *word* embraces at least three basic senses (cf. Brinton 2000, Lipka 1992, Matthews 1991, Mukherjee 2005). The first sense is that of a physical unit with a respective phonological or orthographical representation. The concrete realisation of a word consequently occurs between spaces in writing or between potential pauses in speech (potential in so far as we usually tend to pause between phrases or longer parts of speech rather than after every single word). This definition of wordhood is obviously the one which is considered as the basis for any computer-assisted analysis. The characterisation of a word as a physical entity can pose a potential conflict between the word as a physical unit and the second sense which is that of an abstract unit of meaning. The abstraction is realised by a lexeme, with the help of which a mental concept is related to a given referent. In doing so, a word has a semantic coherence and expresses a more or less codified meaning. A lexeme can be regarded as a vocabulary item or a (mental) dictionary entry, which embraces all inflected forms. A particular word form can thus be the actual manifestation of a lexeme. The word as a unit of meaning is the characterisation that I am interested in as this is the concept that is supposed to be subject to a lexical popularisation process. To a certain extent, this process also takes account of the third sense of wordhood in which the word is regarded as a morphosyntactic unit with an attributed grammatical meaning. Words serve a syntactic purpose, have a grammatical function and fall into particular classes. Internally, a word has a morphological cohesion whereas externally it can be (re-)arranged into different syntactic structures.

Adding a wider sense to the above definitions, Lipka (1992) keeps hold of the extralinguistic reality by means of the concept-forming power of the word. Concepts refer to the semantic dimension of words and are therefore an addition to the above mentioned second sense of wordhood.[8] They target the referential nature of a word and are understood as abstract units of meaning which can be realised by different forms. Meanings can specialise and diverge as words enter into syntagmatic relations with the words that surround them. According to their semantic properties, concepts function as the superordinate link between lexemes in a word family or in a lexical or semantic field. Aspects of lexical semantics will be dicussed in Chapter 5, in which the lexical environment of relevant words is closely examined.

The term *lexical item* has widely been used synonymously with *word* (cf. Carter 1987). Yet, Hoey (1991) and Sinclair (2004) use the term to refer to discrete units that are part of an open class rather than a closed set. They regard the lexical item as a category which is distinct from the word according to the above definitions as it also accounts for multiword expressions, fixed phrases, phrasal verbs etc. With such a categorisation always reflecting group specific patterns, the use of particular lexical items is regarded as a typical characteristic of language variation according to the subject matter or field of discourse. When I use the term *lexical item*, I will thus refer to one or more words considered as a unit, e.g. *Hong Kong*. This will be specified whenever and wherever necessary. In the present computer-assisted analysis, the corpus software processes an n-gram such as *Hong Kong* as separate words, whereas they would semantically be regarded as one expression which is made up of two individual constituents. Counting each word separately is of course due to the fact that the *WordSmith* software identifies a word in the first of the above-mentioned sense which is that of a physically discrete unit. In computer-assisted analysis, the results in the shape of a word are technically regarded as a string of (valid) characters with a given separator to the left and to the right side. Valid characters in *WordSmith* include all letters together with accented and user-defined characters that can be included within a word such as a hyphen or an apostrophe. I have opted for not including such characters in the initial analysis since especially hyphenation is treated ambiguously throughout the corpus texts (e.g. *SARS-virus* or *SARS virus*). However, hyphens will only be ignored in the first part of the analysis, that is, in the predominately quantitative one, since the subsequent qualitative analyses will account for words as complex units of meaning by examining their lexical environment. This also holds true for the treatment of apostrophes, contractions and negatives. Apart from counting and listing words separately, the computer-assisted analysis poses further challenges which one has to be

[8] I am aware that the term *concept* is also treated ambiguously. I will refer to the notion of concept as described by Lipka (1992) and Esser (1993).

aware of before interpreting the data. Should, for example, individual word forms be counted or rather lemmas or word families? According to Nation and Waring (1997) lemmas are defined as a group of inflectional variants of a word, whereas a word family groups together the lemmas and their common derivatives. It is an established thought that lemmas and inflected word forms are bound to share the same meaning and patterns, with patterns being a word or a structure which occurs frequently with the object of enquiry. However, Sinclair (1991) argues that in the light of corpus data this assumption has to be falsified and gives sufficient evidence that each distinct form is potentially a unique meaningful item. Esser (1999) supports Sinclair's view and proves the hypothesis with a collocational analysis of *birthday* and *birthdays*. He shows that the singular form occurs in personal, interactive contexts, whereas the plural form occurs more often in detached, reported contexts. The results of the pilot study are in line with this theory, given that the word forms *patient* and *patients* indeed show different collocational profiles and individual patterns. Additionally, in the absence of (parts-of-speech) tags the software would rather end up counting only a close estimate of lexemes since not all the different senses of word forms would be disambiguated. Taking this into account, I base my analysis on the individual word forms rather than lemmatise the entries. As far as word families are concerned, derivatives will be treated in the same way.

In computer terms it is also difficult to distinguish automatically between homonyms and homographs (e.g. *patient* as a noun or an adjective), polysemic words and, depending on the software, also between upper case and lower case words (e.g. *WHO* and *who*). They can be combined and thus counted as one entry and therefore require manual editing. So, whenever required, I will try to specify the meaning already in the quantitative analysis by examining their lexical environment provided in the concordance lines. As far as the different expressions are concerned, I will henceforth use the term *word* in a general sense and, whenever necessary, make the distinction between *physical unit*, *lexeme*, *word form*, *lexical item* or *concept*.

4.2 Lexical inventories

A study of the frequency distribution of the vocabulary produce some initial empirical findings regarding intuition about some of the lexical differences between the text genres. Lexical constraints often become apparent in word choice, even if the results are evaluated in the absence of an interpretative (lexical) framework. Consequently, before going beyond quantitative patterns in order to give functional interpretations, the description of the individual word choice has to be the first step of characterisation. The frequency of occurrence determines its significance and serves as a basis for explaining why the patterns found exist. Quantitative and statistical findings, i.e. word frequencies, can thus give information about words that are most likely to be used in a certain domain.

4.2.1 Basic wordlists

4.2.1.1 Numerical data

Although the use of quantitative data is often the first stage of corpus-linguistic investigation, it should be clearly set apart from advanced statistical methods used in other (linguistic) disciplines. Nevertheless, by performing even simple numerical analyses, the calculated facts and figures give a first insight into the corpora. Table (13) shows the basic information for both corpora after the computation of a batch word list.[9]

Table 13: *Corpus data*

Unit	Pop. SC	Med. SC
Tokens (running words) in text	297,883	187,080
Tokens used for word list	293,198	177,505
Numbers	4,685	9,575
Types (distinct words)	14,085	9,033
Type/token ratio (TTR)	4.80	5.09
Standardised TTR	45.75	37.38
Standardised TTR basis	1,000	
Mean word length (in characters)	4.86	5.19

The deviance between the total number of tokens and the number of tokens actually used for the word lists is due to the fact that numbers are recognised as tokens in the text but not as valid characters for words. As a consequence, the number of items not used for the word lists is identical with the count of numbers given in Table (13). In this connection, it is of interest that the medical corpus, although one third smaller than the popular one, contains twice as much numerical data. This is in line with the expectations since the scientific texts make regular use of figures as a genre routine (see Chapter 2.3).

[9] The statistics may vary from software to software as they use different criteria for wordhood. All of the results provided hold thus only true for *WordSmith* Version 4. It is well known that even different versions of one and the same software can yield different results (e.g. *WordSmith* Version 3 adds approx. 10,000 words in comparison to Version 4).

As far as types are concerned, the higher number of types in the popular subcorpus has to be related to corpus size. Consequently, the type-token ratio (TTR) can also vary due to the size of the corpus which is being studied. A small corpus might reach a TTR of more than 50%, whereas a larger corpus, containing over a million words, might compute a TTR of below 5%. It is thus only informative and reliable if equal-sized corpora or segments are compared. So the conventional, unstandardised TTR can be – but does not necessarily have to be – misleading. If texts of different lengths are compared, *WordSmith* uses the standardised type-token ratio (STTR) which is defined as follows:[10]

> The standardised type/token ratio is computed every n words as Wordlist goes through each text file. By default, n = 1,000. In other words the ratio is calculated for the first 1,000 running words, then calculated afresh for the next 1,000, and so on to the end of your text or corpus. A running average is computed, which means that you get an average type/token ratio based on consecutive 1,000-word chunks of text. (Scott 1999: *WordSmith* Manual)

The STTR figures shown in Table (13) are consequently percentages of new types for every 1000 tokens. It is interesting to look at the two different ratios and consider how they are possibly related. Specialised language texts are usually expected to have a higher TTR (cf. Bowker and Pearson 2000) and while the conventional TTR seems to be in line with these observations (with a TTR of 4.80 for the popular subcorpus versus a TTR of 5.09 for the medical counterpart), the adjusted STTR yields results contrary to expectations. Whereas the popular subcorpus shows an STTR of 45.75, the medical subcorpus only shows an STTR of 37.38. This seems to be a feature that can be interpreted alongside the difference in size as well as the difference in lexical variation.

In connection to this, Kennedy (1998: 96) states that one of the important outcomes of corpus-based studies is "the recognition that, in whatever textual source, a small number of different word types make up most of the actual word occurrences or tokens". Indeed, most of the words occur only once and are known as *hapax legomena* or *hapaxes* for short. The proportion of hapaxes in the two corpora is consistent with this observation. The complete popular subcorpus consists of 14,085 types and 8,119 of them occur only once. This accounts for 57.6% of hapaxes. The medical subcorpus consists of 9,033 types of which 5,574 occur only once, adding up to a percentage of 61.7% of hapaxes in this collection of texts. This suggests that there is a possible link between the STT ratio and the occurrences of single types. Yet, these ratios can only measure

10 The *WordSmith* method of computing STTR was Mike Scott's invention but parallels a method devised by the mathematicians David Malvern and Brian Richards (University of Reading).

how varied the words in the respective texts are but they can certainly not give any information about what kinds of words they are (e.g. general or specific) or how difficult or rare they are.

The same holds true for mean word length, as it gives only a rough guide to the complexity of the words used. Although the mean word lengths (4.86 characters per word in the popular subcorpus versus 5.19 characters per word in the medical subcorpus) support the view that the texts appear rather similar, the distribution of frequencies for word lengths, as given in Table (14), shows that this holds only partly true since the statistically slight difference in word lengths is due to the complex words used in the medical texts. Although they are not very frequent in a text, complex words are exactly those elements that are perceived as foregrounded items and can thus act as a potential barrier in field-external communication (see Chapter 2.4). Table (14) gives the mean length of words in both subcorpora:

Table 14: *Mean word lengths*

Length	Pop SC	Med SC	Length	Pop SC	Med SC
1-letter words	8,516	7,964	13-letter words	1,058	1,608
2-letter words	46,418	29,102	14-letter words	356	694
3-letter words	53,918	29,168	15-letter words	142	444
4-letter words	51,114	25,803	16-letter words	32	158
5-letter words	32,583	16,854	17-letter words	3	72
6-letter words	25,667	12,065	18-letter words	6	83
7-letter words	25,841	13,232	19-letter words	0	41
8-letter words	19,498	15,466	20-letter words	0	15
9-letter words	14,473	10,008	21-letter words	0	3
10-letter words	8,133	7,051	22-letter words	0	2
11-letter words	4,608	5,235	23-letter words	0	5
12-letter words	2,221	2,423	24-letter words	0	6

The figures show that there is first of all a similarity in the distribution of words with a moderate length as both corpora peak in 3-letter words. The rankings are almost identical for the types that contain up to 10 letters but start to diverge from that word size onwards. The medical subcorpus contains a significant variety of highly complex words that consist of up to 24 letters and are not used in the popular representation of SARS. This includes terms such as *electrocardiographic* or *microimmunofluorescence.* As explained in Chapter 2, the terms are fairly complex constructions but they are transparent to the scientific audience as they deduce the specific meaning from the component morphemes. Niederhauser (2001) claims that the reduction of morphosyntactic complexity is one of the basic processes of popularisation and the results seem to confirm this hypothesis. Yet, the question arises whether the terms are just eliminated from the text or whether they undergo a process of transfer. This will be put to the test in the qualitative analysis. To compare the two corpora on the lexical level, it is first of all necessary to retrieve relevant units and ideally candidate patterns emerging out of them.

4.2.1.2 Basic frequency rankings

The production of word lists is a routine procedure to explore a corpus. It is a plain but nevertheless powerful tool since the frequency of occurrence helps to determine the role of the respective words. Furthermore, it helps to obtain a reliable description of regularities in the texts as it focuses on repeated events rather than on the individual meaning or function. Tables (15) and (16) give a basic ranking of all types according to frequency of occurrence for the most common words in the subcorpora, with a cutting-off point at 100 entries.[11] Although the pilot study has shown that it is more useful to take the refined lists as a basis for analysis rather than the basic one, there are several points of interest within the raw lists that I would like to comment on before proceeding. In addition, there are items in these lists that should be manually removed. Although numbers (indicated by #) and clitics (*S*, *T*) are prominent items as far as frequency is concerned, they have to be excluded in the refined lists since they cannot be counted as words.

The greatest part of types in the basic top 100 ranking are function words. According to Brinton (2000:118), they are closed classes, represent the minor parts of speech and thus carry less of the communicative force of a text. They are nevertheless crucial to the texts as they express grammatical meaning by relating sentence parts to each other. The prominent function words in the two corpora are, according to frequency, definite and indefinite articles (*the*, *a*), prepositions (*to*, *of*, *in*, *on*, *at*, *with*, *for*, *from*, *by*, *up*), conjunctions (*and*, *as*, *but*, *because*, *also*, *if*), auxiliary verbs (forms of *to have* and *to be*), personal pronouns (*he*, *she*, *I*, *we*, *they)*, and demonstrative or relative pronouns (*that*, *it*). The use of personal pronouns differs in the two collections of text, as the medical wordlist includes only the plural forms *we*, *they* and *our*, whereas in the popular texts the first and third person singular is used, i.e. *I*, *he*, *she*. This does not come as a surprise since personalisation is an important strategy of the news media whereas it is generally avoided in scientific publications. The pronouns

[11] This decision is based on Kennedy's findings (1998:69), who has shown that in a corpus between 50 and 100 types typically account for up to half of the total tokens in a text (this is known as *Zipf's law*).

that are used reflect the writing convention as multiple authorship is common in scientific publications. The distribution of modal verbs also differs in the two subcorpora. The popular subcorpus shows more modals in high frequency (*may*, *can*, *would*, *could*), whereas the medical counterpart makes only use of *may* and *should* and they rank significantly lower.

Table 15: *Basic frequency ranking for the popular subcorpus*[12]

N	Word	N	Word	N	Word	N	Word
1	THE	26	HAS	51	ONE	76	OUT
2	TO	27	DISEASE	52	WHICH	77	SO
3	OF	28	BEEN	53	THAN	78	FIRST
4	IN	29	PEOPLE	54	ABOUT	79	SOME
5	A	30	BUT	55	WORLD	80	AFTER
6	AND	31	VIRUS	56	THEIR	81	WHEN
7	#	32	HAD	57	PATIENTS	82	REPORTED
8	THAT	33	BY	58	ITS	83	NOW
9	SARS	34	CASES	59	WILL	84	CHINESE
10	IS	35	CHINA	60	NO	85	UP
11	S	36	HONG	61	SPREAD	86	LAST
12	IT	37	KONG	62	OTHER	87	HIS
13	HAVE	38	THEY	63	T	88	PUBLIC
14	FOR	39	NOT	64	INFECTED	89	WHERE
15	SAID	40	WE	65	MAY	90	ALSO
16	WAS	41	AN	66	IF	91	MOST
17	WHO	42	WERE	67	OUTBREAK	92	RESPIRATORY
18	FROM	43	HE	68	CAN	93	SYMPTOMS
19	ON	44	THIS	69	BEIJING	94	TORONTO
20	WITH	45	MORE	70	WOULD	95	DOCTORS
21	BE	46	THERE	71	COULD	96	MANY
22	HEALTH	47	OR	72	I	97	BECAUSE
23	ARE	48	NEW	73	GOVERNMENT	98	SHE
24	AT	49	OFFICIALS	74	ALL	99	INTO
25	AS	50	HOSPITAL	75	TWO	100	DAYS

According to Brinton (2000:118), content words carry the primary communicative force as they have referential or cognitive meaning, are part of an open class and variable in form. *Content* or *lexical words* represent the major parts of speech and include nouns, verbs, adjectives, adverbs and some pronouns. The first word in both lists which contributes lexical meaning is the name of the disease in the form of its eminent acronym SARS. Although it is the

12 Detailed frequency data for each of the following lists, indicating number of tokens, percentages and numbers of texts in which the word occurs, are given in the Appendix.

agreed-on denomination of the disease, it is nevertheless noteworthy to observe that the acronym is institutionalised in the popular representation of the disease, whereas the component words themselves which contribute their initials (*S*evere *A*cute *R*espiratory *S*yndrome) are not consistently among the one hundred most frequent words. In the popular subcorpus, only *respiratory* can be found at rank 92. In the medical subcorpus, all components of the acronym are among the top 100 words, although their distribution differs as *respiratory* holds rank 31, whereas *severe*, *acute*, and *syndrome* rank significantly lower (in position 56, 62 and 67 respectively). It is interesting to find out if this is consistent feature that will turn out as salient in the course of the study

Table 16: *Basic frequency ranking for the medical subcorpus*

N	Word	N	Word	N	Word	N	Word
1	#	26	HEALTH	51	CLINICAL	76	WHICH
2	THE	27	AT	52	AFTER	77	CONTROL
3	OF	28	ARE	53	S	78	ASSOCIATED
4	AND	29	THIS	54	OUTBREAK	79	NEW
5	IN	30	BEEN	55	PROBABLE	80	TWO
6	TO	31	RESPIRATORY	56	SEVERE	81	ILLNESS
7	SARS	32	PATIENT	57	TRANSMISSION	82	CHINA
8	A	33	AN	58	IT	83	FEVER
9	WITH	34	VIRUS	59	ONE	84	VIRAL
10	FOR	35	HAS	60	CONTACT	85	SUSPECTED
11	PATIENTS	36	NOT	61	COV	86	TORONTO
12	WERE	37	CASE	62	ACUTE	87	HUMAN
13	WAS	38	CORONAVIRUS	63	MORE	88	INFECTED
14	ON	39	HOSPITAL	64	MAY	89	TRAVEL
15	THAT	40	CARE	65	ALSO	90	FIRST
16	IS	41	WE	66	NO	91	THEY
17	FROM	42	DISEASE	67	SYNDROME	92	USED
18	CASES	43	ALL	68	BUT	93	APRIL
19	WHO	44	DAYS	69	OUR	94	PNEUMONIA
20	OR	45	OTHER	70	SYMPTOMS	95	TIME
21	HAVE	46	INFECTION	71	THERE	96	MOST
22	BY	47	THESE	72	MARCH	97	CHEST
23	AS	48	REPORTED	73	THEIR	98	SHOULD
24	BE	49	HONG	74	DURING	99	IDENTIFIED
25	HAD	50	KONG	75	THAN	100	ONLY

In the following, I will take a brief account of prominent content words, first in the wordlist for the popular texts. The by far dominating word class is that of nouns, followed by adjectives and adverbs. There is already a prominent group of frequent nouns which can be grouped thematically together into those which belong to the semantic field of (1) medical features of SARS; (2) affected persons and people involved; and (3) locative or proper nouns that refer to affected areas. The first group of nouns is the one that is most likely to be related to the medical domain. *Health* is the second most frequent lexical word, followed by the antonym *disease*. The agent of the disease is subsequently given, which is a *virus* that infects *people* who show *symptoms* and are admitted to a *hospital*. Related to this, SARS *spreads* and thus causes an *outbreak*. The second prominent group of nouns relates to persons affected by or involved in the epidemic. *People* and *public* are the most general realisation of this recurring concept but it is also specified by referring to those who have SARS as *patients*, or more depersonalised, as *cases*. Professionals involved are *doctors* who take care of the patients and those who are concerned in the management of the disease, i.e. *officials* and *government*. The third group of nouns includes proper names of countries and cities that refer to the affected areas such as *China*, *Hong Kong*, *Beijing* and *Toronto*. As far as verbs, adjectives and adverbs are concerned, the types *public* and *reported* are examples that illustrate the aforementioned drawback of single wordlists. *Public* can either be a noun, referring to the *general public*, or an adjective as in, for example, *public concern*. Likewise, *reported* is either used as an inflected verb to quote somebody who has reported on something, or as an adjective as in *reported SARS cases*; the same holds true also for *infected*.

As far as terminology is concerned, the content words in the popular wordlist can be regarded as general rather than specific since only *respiratory* remains as a set medical term. The other words that are derived from the medical domain, such as *virus* and *disease*, are expressions that have already been institutionalised in general language use and are thus called semi-technical words (cf. Busch 1994, Fischer 1998). As Table (16) shows, this is obviously different in the medical subcorpus.

As far as the content words are concerned, the dominating word class in the medical texts is again nouns. They can also be grouped into preliminary fuzzy categories that overlap with those defined for the popular subcorpus. The concept of affected persons is realised again, however, not only the plural for *patients* and *cases* is given but also their singular form. *Disease* is also among the most frequent content words but does not have such a prominent position in terms of frequency. In this connection, SARS is furthermore referred to as *pneumonia*, *illness* or *infection* and the causative agent is not only given as *virus* but, more specifically, as *coronavirus* or *COV* for short. There is a group of words that is related to the spread and containment of the disease (*transmission*, *contact*, *outbreak*, *control*) and the affected areas are presented as well (*China*, *Hong Kong*, *Toronto*). Contrary to the popular corpus, there are also types referring to time (*days*, *March*, *April*). As far as other word classes are concerned, there are more adjectives and adverbs in the medical wordlist since *respiratory* is shared by both subcorpora but adjectives such as *clinical*, *probable*, *severe*, *viral* and *human* are, for now, specific to the medical representation. This does not come as a surprise as some of the most important components of LSP are conceptual units that are expressed as nominal groups with nouns being the *determinatum* and adjectives being the *determinans*. Denominations, such as *Severe Acute Respiratory Syndrome*, illustrate that it is this very combination of items that leads to the detailed description of a disease or process required in scientific discourse.

Another prominent, genre-related feature is the top position of numbers in the medical texts indicated by the hash symbol #. According to the statistics, there are 9,588 figures in total, adding up to a percentage of 5.12, which has been counted in 98.58 percent of the medical texts. This is, as already outlined in Chapter 2, certainly due to the nature of scientific publications which present results of studies and clinical trials with numerical background information. In connection to this, it is interesting to investigate if the prominent number of comparative adjectives and adverbs in the popular subcorpus, i.e. relational quantifiers such as *more*, *some*, *many*, *most*, *other* and *all*, can be seen in relation to this.

4.2.2 Refined wordlists

4.2.2.1 The *stoplist* function

The comparison of the two basic wordlists shows that many general function words are used frequently in both genres. This is in line with observations that on statistical grounds a prominent number of function words are frequently used across a range of genres. The conception of disposability in a wide range of domains indicates that they fulfil primarily syntactic functions. It is thus a rational for excluding these common (function) words from the detailed analysis, as I suppose that they are neither genre- nor topic-specific. As I am interested in content words as units of meaning in accordance with the definition given earlier, it is thus useful to employ a *stoplist* that excludes most of the common words from the basic wordlists. The pilot study has proven that this is a step worth taking and that the results obtained will provide a valuable basis for the subsequent lexical analysis. As explained in the previous Chapter, a stoplist contains a record of words which are not to be included in the analysis. To generate such a list, either manually selected items have to be included in a plain text file, or lists specifying all the words one wishes to ignore can be exported. I am opting for the latter, as using items from an existing general language corpus ensures an unbiased selection.[13] To take account of the larger size of the two subcorpora compared to the pilot study, I have decided on producing a stoplist of the 100 most frequent words in the FLOB corpus instead of the previously used 50 most frequent ones. Table (17) therefore gives the top 100 entries in order of frequency for the FLOB corpus that will be excluded from the analyses of the two subcorpora:

Table 17: *Top 100 entries in the FLOB corpus*

N	Word	N	Word	N	Word	N	Word	N	Word
1	THE	21	HAD	41	THEIR	61	THEM	81	VERY
2	OF	22	THIS	42	HAS	62	CAN	82	OUR
3	AND	23	NOT	43	WOULD	63	INTO	83	WELL
4	TO	24	BUT	44	WHEN	64	THAN	84	NEW
5	A	25	FROM	45	IF	65	TWO	85	AFTER
6	IN	26	HAVE	46	SO	66	COULD	86	SUCH
7	THAT	27	ARE	47	NO	67	ME	87	MUST
8	IS	28	WHICH	48	HIM	68	THEN	88	MAN
9	WAS	29	HER	49	WILL	69	OTHER	89	MUCH
10	IT	30	SHE	50	WHO	70	ITS	90	YEARS
11	FOR	31	OR	51	MORE	71	THESE	91	MOST
12	HE	32	YOU	52	OUT	72	DO	92	BEFORE
13	AS	33	THEY	53	SAID	73	NOW	93	MANY
14	BE	34	AN	54	UP	74	MAY	94	WHERE
15	WITH	35	WERE	55	ABOUT	75	ANY	95	EVEN
16	ON	36	ONE	56	SOME	76	FIRST	96	ALSO
17	I	37	BEEN	57	WHAT	77	OVER	97	BACK
18	HIS	38	THERE	58	ONLY	78	MADE	98	WAY
19	AT	39	ALL	59	MY	79	SHOULD	99	THOSE
20	BY	40	WE	60	TIME	80	LIKE	100	BEING

The list shows that in a general, balanced corpus (as opposed to a specialised or genre-restricted corpus) the vast majority of the words are function words. Even when the 100 most frequent words are considered, instead of the 50 used previously in the pilot study, 97 percent of them are found to be function words again, with the dominant word classes being determiners, conjunctions, prepositions, pronouns and auxiliaries. Focusing on rank order, the results of the

[13] Whenever a general reference corpus is needed in the course of the analysis, I will refer to the FLOB-corpus. For a description of the corpus see 3.4.

three wordlists show a remarkable degree of consistency since the top positions are identical. It is thus evident that they are disposable across a variety of genres and registers. Nevertheless, the study of the basic frequency lists has shown that the more specialised or homogeneous the corpus is the more content words there will be among the top most frequent words.

Apart from exporting the words extracted from the FLOB wordlist into the stoplist file, it will also be manually edited in order to exclude the aforementioned numbers and clitic forms produced by *WordSmith*. As a consequence, the results will be manually edited in the refined wordlists.

4.2.2.2 Refined frequency rankings

Tables (18) and (19) give the refined ranking for the 100 most frequent words in both subcorpora.

Table 18: *Refined frequency ranking for the popular subcorpus*

N	Word	N	Word	N	Word	N	Word
1	SARS	26	DOCTORS	51	HOSPITALS	76	WHILE
2	HEALTH	27	BECAUSE	52	TODAY	77	AROUND
3	DISEASE	28	DAYS	53	COUNTRIES	78	UNDER
4	PEOPLE	29	WEEK	54	INFECTION	79	DISEASES
5	VIRUS	30	EPIDEMIC	55	SCIENTISTS	80	CARE
6	CASES	31	HOW	56	ASIA	81	INFECTIOUS
7	CHINA	32	TRAVEL	57	CANADA	82	MIGHT
8	HONG	33	SINGAPORE	58	ILLNESS	83	PATIENT
9	KONG	34	CITY	59	DAY	84	GLOBAL
10	OFFICIALS	35	SAYS	60	SYNDROME	85	INCLUDING
11	HOSPITAL	36	MEDICAL	61	LEAST	86	INFORMATION
12	WORLD	37	CONTROL	62	QUARANTINE	87	US
13	PATIENTS	38	DEATHS	63	PROVINCE	88	UNIVERSITY
14	SPREAD	39	YESTERDAY	64	ACUTE	89	DR
15	INFECTED	40	YEAR	65	FAR	90	SINCE
16	OUTBREAK	41	GUANGDONG	66	HOME	91	TAIWAN
17	BEIJING	42	STILL	67	CONTACT	92	MARCH
18	GOVERNMENT	43	SUSPECTED	68	COUNTRY	93	DOWN
19	REPORTED	44	CASE	69	AUTHORITIES	94	WEEKS
20	CHINESE	45	DIED	70	ANOTHER	95	MASKS
21	LAST	46	SEVERE	71	FLU	96	KNOWN
22	PUBLIC	47	NUMBER	72	AGAINST	97	TOLD
23	RESPIRATORY	48	THREE	73	DEATH	98	SAY
24	SYMPTOMS	49	WORKERS	74	PNEUMONIA	99	CORONAVIRUS
25	TORONTO	50	JUST	75	THROUGH	100	DIRECTOR

After recomputing the popular wordlist, 132,882 tokens (function words and odd letters) have been removed in the popular and 76,754 tokens in the medical subcorpus. This difference is first of all due to the overall frequency in size as the medical subcorpus is a third smaller, but is also accounted for by the density of the medical texts as far as content words are concerned.

Table 19: *Refined frequency ranking for the medical subcorpus*

N	Word	N	Word	N	Word	N	Word
1	SARS	26	SYNDROME	51	NUMBER	76	STATES
2	PATIENTS	27	SYMPTOMS	52	PERSONS	77	UNITED
3	CASES	28	MARCH	53	DATA	78	DEFINITION
4	HEALTH	29	DURING	54	SPREAD	79	INFORMATION
5	RESPIRATORY	30	CONTROL	55	INCLUDING	80	MEDICAL
6	PATIENT	31	ASSOCIATED	56	WORKERS	81	SPECIFIC
7	VIRUS	32	ILLNESS	57	PCR	82	ISOLATION
8	CASE	33	CHINA	58	HOWEVER	83	ANALYSIS
9	CORONAVIRUS	34	FEVER	59	ONSET	84	AVAILABLE
10	HOSPITAL	35	VIRAL	60	THREE	85	WORLD
11	CARE	36	SUSPECTED	61	SAMPLES	86	AREAS
12	DISEASE	37	TORONTO	62	CDC	87	KNOWN
13	DAYS	38	HUMAN	63	DAY	88	VENTILATION
14	INFECTION	39	INFECTED	64	FIGURE	89	EXPOSURE
15	REPORTED	40	TRAVEL	65	INFECTIOUS	90	SERUM
16	HONG	41	USED	66	TREATMENT	91	QUARANTINE
17	KONG	42	APRIL	67	RIBAVIRIN	92	HOSPITALS
18	CLINICAL	43	PNEUMONIA	68	USE	93	WITHIN
19	OUTBREAK	44	CHEST	69	C	94	ICU
20	PROBABLE	45	IDENTIFIED	70	BOTH	95	DEATHS
21	SEVERE	46	LABORATORY	71	ALTHOUGH	96	STAFF
22	TRANSMISSION	47	AMONG	72	MIGHT	97	ADMISSION
23	CONTACT	48	PUBLIC	73	BECAUSE	98	EARLY
24	COV	49	HIGH	74	INDEX	99	ILL
25	ACUTE	50	CELLS	75	BETWEEN	100	MEASURES

Since there are now hardly any function words left, this procedure shows nicely that there is by and large an agreement in general words between the restricted SARS corpus and the homogenous FLOB corpus. Since only the most frequent content words are displayed, the stoplist is certainly a step worth taking as the lists now present a valid snapshot of the 'aboutness' of the texts.14 The refined lists give a good indication of the nature of the texts as they include almost

[14] I have conducted an alphabetical case sensitive search for *WHO* (organisation vs. pronoun), *US* (abbreviation vs. pronoun), and *PATIENT* (noun vs. adjective), to make a precise distinction between homographs/homonyms. The results will be considered in the course of the analysis.

exclusively lexical words. The function words left, such as the conjunction because or the modal might, are for the most part shared by both lists. What is now obvious is that both subcorpora not only share a prominent group of function words but that there is also some lexical overlap in regard to content words. I will now take the refined lists as a basis for a detailed study of single words that are shared by both subcorpora and of those that are used frequently only in one collection of texts.

4.3 Contrasting the lexical inventories

4.3.1 Words in common

The pilot study (see Chapter 3.4) has already shown that one third of the types in the frequency lists – for the top 50 items – are shared by both subcorpora and that they have a common stock of words. The large-scale analysis for the full corpus not only confirms these early observations but shows that almost half of the types occur in both corpora among the 100 most frequent items. The results of the comparison are summarised in Table (20). The first column gives the word that is shared and the following two columns give the frequencies and numbers of words, ordered according to the popular subcorpus.

Table (20) shows that 49 types out of 100 are shared but it has to be pointed out that this agreement might be in form only as the shared words may function differently in the two subcorpora. I am aware that the shared words are, like all language items, tied to their situational context and verbal co-text that affect their meaning as well as their function. However, the comparison of single words can already provide insights into the style and 'aboutness' of the two collections of texts.

The list reveals a consistency not only in the frequent use of general words, such as *world*, but moreover in the use of semi-technical words that are institutionalised in every-day language use, such as *symptoms*, and there is also a small amount of set medical terms, for example *pneumonia* and *coronavirus*. The overall presence of semi-technical expressions in both genres suggests that the common vocabulary gradually adapts to the domains. Items that are transferred literally into the popular domain are actually those that are simple in nature and are not foregrounded in general language, yet reflect the topic suitably. It is of interest for the qualitative analysis whether the remaining scientific terms are used in a different way to bridge this gap.

Table 20: *Words in common*

N	Shared Word	Rank (Freq.) Pop C.	/ Med C.	N	Shared Word	Rank (Freq.) Pop C.	/ Med C.
1	SARS	1 (1.80%)	1 (1.61%)	25	DEATHS	38 (0.11%)	95 (0.07%)
2	HEALTH	2 (0.58%)	4 (0.41%)	26	SUSPECTED	43 (0.11%)	36 (0.12%)
3	DISEASE	3 (0.51%)	12 (0.28%)	27	CASE	44 (0.10%)	8 (0.29%)
4	VIRUS	5 (0.44%)	7 (0.32%)	28	SEVERE	46 (0.10%)	21 (0.18%)
5	CASES	6 (0.41%)	3 (0.57%)	29	NUMBER	47 (0.10%)	51 (0.10%)
6	CHINA	7 (0.40%)	33 (0.13%)	30	THREE	48 (0.10%)	60 (0.10%)
7a	HONG KONG	8 & 9	16 & 17a	31	WORKERS	49 (0.10%)	56 (0.10%)
7b		(0.39%)	(0.23%)	32	HOSPITALS	51 (0.10%)	92 (0.08%)
8	HOSPITAL	11 (0.27%)	12 (0.29%)	33	INFECTION	54 (0.09%)	14 (0.25%)
9	WORLD	12a (0.23%)	85 (0.08%)	34	ILLNESS	58 (0.09%)	32 (0.14%)
10	WHO	12b (0.23%)	17b (0.22%)	35	DAY	59 (0.09%)	63 (0.09%)
11	PATIENTS	13 (0.22%)	2 (0.94%)	36	SYNDROME	60 (0.09%)	26 (0.16%)
12	SPREAD	14 (0.22%)	54 (0.10%)	37	QUARAN-TINE	62 (0.09%)	91 (0.08%)
13	INFECTED	15 (0.20%)	39 (0.12%)	38	ACUTE	64 (0.09%)	25 (0.17%)
14	OUTBREAK	16 (0.19%)	19 (0.19%)	39	CONTACT	67 (0.09%)	23 (0.17%)
15	REPORTED	19 (0.16%)	15 (0.23%)	40	PNEUMONIA	74 (0.08%)	43 (0.11%)
16	PUBLIC	22 (0.15%)	48 (0.10%)	41	CARE	80 (0.08%)	11 (0.29%)
17	RESPIRA-TORY	23 (0.14%)	5 (0.36%)	42	INFECTIOUS	81 (0.07%)	65 (0.09%)
18	SYMPTOMS	24 (0.14%)	27 (0.16%)	43	MIGHT	82 (0.07%)	72 (0.08%)
19	TORONTO	25 (0.13%)	37 (0.12%)	44	PATIENT	83 (0.08%)	6 (0.35%)
20	BECAUSE	27 (0.13%)	73 (0.08%)	45	INCLUDING	85 (0.07%)	55 (0.10%)
21	DAYS	28 (0.12%)	13 (0.27)	46	INFORMA-TION	86 (0.07%)	79 (0.08%)
22	TRAVEL	32 (0.12%)	40 (0.11%)	47	MARCH	92 (0.07%)	28 (0.16%)
23	MEDICAL	36 (0.11%)	80 (0.08%)	48	KNOWN	96 (0.07%)	87 (0.08%)
24	CONTROL	37 (0.11%)	30 (0.15%)	49	CORONA-VIRUS	99 (0.06%)	9 (0.29%)

It can be seen that the words shared by the two subcorpora fall, to a large degree, into a limited number of recurring semantic fields. These semantic groups span lexical word classes and present a valid snapshot of the subject matter. Four main fuzzy categories can be established: (1) *affected persons and institutions*; (2) *place and time reference*; (3) *medical features of SARS*; and (4) *outcome and measures*.

(1) 'persons & institutions': people that are directly affected by SARS or organisations that are involved in the management of the disease represent a small but distinctly recognisable semantic group. Within this group, a demarcation can be made between those that have fallen ill, i.e. *patient(s)* and *cases(s)*, those that are indirectly affected, i.e. *public* and *workers*, and the institutions that are involved in the treatment and control of infections, i.e. *hospital* and *WHO*.

(2) 'place & time': the areas where the virus has spread from, i.e. *China*, where the cases have peaked, i.e. *Hong Kong*, and where it has spread to in the *world* via *travel*, e.g. *Toronto*, are included in this category. The proper names refer to the key places where SARS has hit worst, and the date *March* points out when it hit.

(3) 'medical features of SARS': this is the group that contains only semi-technical terms and the few restricted medical terms. SARS is referred to as an *illness* or *disease*, along the same line as *pneumonia*, and the cause for the *infection* is given as *virus*, or more specifically, as *coronavirus*. SARS is a *respiratory* disease and the people that are infected have particular *symptoms* and are thus *suspected* cases. A popularisation process should take place predominantly within this category because it is evidently subject- rather than genre-specific.

(4) 'outcome & measures': there is also a distinct group of words that describe the effects of SARS as well as possible countermeasures. SARS *spreads* via *contact*, which leads to an *outbreak* of the disease which might result in *deaths*. To prevent effects, the disease needs to be contained by *control* which can, for example, be achieved by establishing *quarantine* units. This is certainly a key issue that is equally important to the general public and medical professionals.

The semantic categorisation does not only illustrate the thematic overlap, but, furthermore, it shows that, on a gradient from everyday language to terminology, both discourse types meet halfway whenever general features of the disease are discussed. It also points out that scientific language does not exist as a separate entity of its own using entirely specific lexis, but that it also employs general language items which relate to external aspects. Nevertheless, LSP research (cf. Sager et al. 1980) gives evidence that general words can also be subject to specification as they are embedded in a specific co-text that alters and narrows their meaning. So, as already pointed out, the results show that there are relationships between the two genres, but they do not shed any light on whether those similarities are in form only or whether they show agreement in their function and use as well.

4.3.2 Words apart

The preceding analysis has supported the view that there are obvious lexical relations and overlaps between the popular and the scientific representation of SARS. Nonetheless, the topic is still a different issue for health professionals and for the general public as they have different interests. While medical scientists focus on the research results and their practical implication, the coverage in the newspress is certainly influenced by the potential for drama as an epidemic poses a risk to the general public. As a result, the analysis of the refined frequency lists has not only revealed a common stock of 49 words, but has also yielded 51 words that are frequent only in one of the corpora. I will now discuss the individual lists in some detail, beginning with the words that are unique to the popular subcorpus. Table (21) gives the items in frequency order, together with their rank in the refined list and their relative frequency:

Table 21: *High frequency words unique to the popular subcorpus*

N	Word	Rank (Freq.)	N	Word	Rank (Freq.)
1	PEOPLE	4 (0.46%)	27	FAR	65 (0.09%)
2	OFFICIALS	10 (0.28%)	28	HOME	66 (0.09%)
3	BEIJING	17 (0.19%)	29	COUNTRY	68 (0.08%)
4	GOVERNMENT	18 (0.18%)	30	AUTHORITIES	69 (0.08%)
5	CHINESE	20 (0.16%)	31	ANOTHER	70 (0.08%)
6	LAST	21 (0.15%)	32	FLU	71 (0.08%)
7	DOCTORS	26 (0.13%)	33	AGAINST	72 (0.08%)
8	WEEK	29 (0.12%)	34	DEATH	73 (0.08%)
9	EPIDEMIC	30 (0.12%)	35	THROUGH	75 (0.08%)
10	HOW	31 (0.12%)	36	WHILE	76 (0.08%)
11	SINGAPORE	33 (0.12%)	37	AROUND	77 (0.08%)
12	CITY	34 (0.12%)	38	UNDER	78 (0.08%)
13	SAYS	35 (0.12%)	39	DISEASES	79 (0.08%)
14	YESTERDAY	39 (0.11%)	40	GLOBAL	84 (0.07%)
15	YEAR	40 (0.11%)	41	US	87 (0.07%)
16	GUANGDONG	41 (0.11%)	42	UNIVERSITY	88 (0.07%)
17	STILL	42 (0.11%)	43	DR	89 (0.07%)
18	DIED	45 (0.10%)	44	SINCE	90 (0.07%)
19	JUST	50 (0.10%)	45	TAIWAN	91 (0.07%)
20	TODAY	52 (0.10%)	46	DOWN	93 (0.07%)
21	COUNTRIES	53 (0.10%)	47	WEEKS	94 (0.07%)
22	SCIENTISTS	55 (0.09%)	48	MASKS	95 (0.07%)
23	ASIA	56 (0.09%)	49	TOLD	97 (0.07%)
24	CANADA	57 (0.09%)	50	SAY	98 (0.07%)
25	LEAST	61 (0.09%)	51	DIRECTOR	100 (0.06%)
26	PROVINCE	63 (0.09%)			

As Table (21) shows, the types that are unique to the popular presentation of SARS display a remarkable focus on the first and on the second semantic category that have been established in the previous section (*'persons & institutions'* and *'place & time'*). In accordance with this, the nature of the words is general as opposed to scientific and just three words, i.e. *flu*, *epidemic* and the plural form *diseases*, fall into category three (medical features). Contrary to the shared word list, the structure is less nominal and there is a new, substantial group of prepositions and quantifiers, such as *under* and *just*. So, there is just one new semantic aspect within the individual words while the categories already established are both remarkably extended and specified:

(1) 'persons & institutions': the first and the second word that are unique to the popular subcorpus, *people* and *officials*, are good examples of the particular reference that is paid to representatives of the scientific community. Among the 51 items there are eleven that refer to this concept and their role is further specified. Medical professionals involved in the topic are referred to as *scientists* or *doctors* (*Dr.*) and the *officials* and *authorities* include personal reference, i.e. *director*, but also collective bodies such as *government* or *university*. The prominence of this concept can be interpreted in relation to the dominant view of science popularisation, as discussed in Chapter 3, which assumes that information is transferred linearly from one discourse to another by means of reported events. There is a basic assumption that goes along with this view, claiming that scientists and scientific institutions are required as authorities on scientific topics because reality is constituted around quotes and references. The verbs that are unique to the subcorpus, i.e. *say(s)* and *told*, support this view.

(2) 'place & time': there are 16 words in total that can be grouped into this category. The group is extended, by including words such as *global*, *Asia* and *country*, but at the same time specified since not only countries such as *Canada* and *Taiwan* are given, but also *provinces*, i.e. *Guangdong*. Note the semantic contrast between *far* and *home*. This opposition illustrates the nature of geographical reference nicely since SARS has its hub in Asia but causes global concern as it also affects, for example, Canada. The preponderance of this concept can be explained through its usefulness for signalling the newsworthiness of the epidemic because it raises the awareness that SARS is a foreign disease that may now be lingering at the doorstep. As far as the time deixis is concerned, the span is also both broadened and narrowed by words such as *week(s)*, *yesterday* and *today*.

(3) 'prepositions & quantifiers': there is a minor group of some prominent prepositions that can be regarded as tangible as opposed to abstract, i.e. *last*, *least* and *just*, but there are also those that give vague reference, such as *around* and *under*, if they are used as quantifiers.

The study of the basic statistics for the individual subcorpora in section 4.2 has shown that a considerably lower amount of numbers and figures is used in the popular than in the scientific texts. The reduction of the amount of numerical data can go hand in hand with the frequent use of verbal quantifiers as the degree of complexity is lowered where accuracy is not required. Furthermore, this 'ready-made' interpretation could prevent a possible misconception by the general audience. All in all, the words that are unique to the popular texts are general in nature, yet proper nouns as well as prepositions and verbs seem to fulfil a certain function. At the same time, however, general lexis does not point to the fact that the popular representation of SARS should only be considered in these limited terms, but that it should be linked to the shared specific reference.

Table 22: *High frequency words unique to the medical subcorpus*

N	Word	Rank (Freq.)	N	Word	Rank (Freq.)
1	CLINICAL	18 (0.20%)	**27**	RIBAVIRIN	67 (0.09%)
2	PROBABLE	20 (0.19%)	**28**	USE	68 (0.09%)
3	TRANSMISSION	22 (0.18%)	**29**	C	69 (0.09%)
4	COV	24 (0.17%)	**30**	BOTH	70 (0.09%)
5	DURING	29 (0.16%)	**31**	ALTHOUGH	71 (0.08%)
6	ASSOCIATED	31 (0.15%)	**32**	INDEX	74 (0.08%)
7	FEVER	34 (0.13%)	**33**	BETWEEN	75 (0.08%)
8	VIRAL	35 (0.12%)	**34**	STATES	76 (0.08%)
9	HUMAN	38 (0.12%)	**35**	UNITED	77 (0.08%)
10	USED	41 (0.11%)	**36**	DEFINITION	78 (0.08%)
11	APRIL	42 (0.11%)	**37**	SPECIFIC	81 (0.08%)
12	CHEST	44 (0.11%)	**38**	ISOLATION	82 (0.08%)
13	IDENTIFIED	45 (0.11%)	**39**	ANALYSIS	83 (0.08%)
14	LABORATORY	46 (0.10%)	**40**	AVAILABLE	84 (0.08%)
15	AMONG	47 (0.10%)	**41**	AREAS	86 (0.08%)
16	HIGH	49 (0.10%)	**42**	VENTILATION	88 (0.08%)
17	CELLS	50 (0.10%)	**43**	EXPOSURE	89 (0.08%)
18	PERSONS	52 (0.10%)	**44**	SERUM	90 (0.08%)
19	DATA	53 (0.10%)	**45**	WITHIN	93 (0.08%)
20	PCR	57 (0.10%)	**46**	ICU	94 (0.07%)
21	HOWEVER	58 (0.10%)	**47**	STAFF	96 (0.07%)
22	ONSET	59 (0.10%)	**48**	ADMISSION	97 (0.07%)
23	SAMPLES	61 (0.10%)	**49**	EARLY	98 (0.07%)
24	CDC	62 (0.09%)	**50**	ILL	99 (0.07%)
25	FIGURE	64 (0.09%)	**51**	MEASURES	100 (0.07%)
26	TREATMENT	66 (0.09%)			

The words that are unique to the medical subcorpus show, as expected, a higher degree of specialisation and semantic variation, although it is not possible to narrow down their meaning easily to certain groups. Table (22) gives the 51 words that are frequently found in the medical subcorpus only.

Contrary to the individual words, the fuzzy semantic groups that were established earlier cannot be applied straightforwardly to the unique medical words. There are only three references to the '*persons & institution*' category, i.e. *persons*, *staff* and *CDC*, and to the 'place and time' category, i.e. *United States*, *April* and *early*. The remaining two categories for 'medical features' and 'effects' are definitely too broad to provide an accurate account of the aboutness of the words in the list. The words are more variable and not as homogenous and fitting them into set categories would not be very useful. Nevertheless, there are at least some salient and recurrent themes that link certain words in the list. The most prominent expressions are those concerned with virus identification: *cells* and *specimens* are *sampled* and a *PCR* test is conducted in a *laboratory* setting to identify the coronavirus or *COV*. Then, there are terms that relate to the care and treatment of infected people, as *measures* include *admission*, *isolation*, *ventilation* and the drug *Ribavirin*. Finally, there is a group of adjectives, such as *clinical*, *probable* and *associated* that seems to reflect the description of cases.
On the whole, these particular issues are no doubt irrelevant to a general audience and the study of the individual wordlists suggests that the lexical differences between the two genres are not necessarily due to the opposition of scientificness vs. generality, but are rather functionally motivated. The overall comparison of the refined wordlists for the popular and the medical subcorpus has shown both a remarkable consistency in terms of shared general and semi-technical words and systematic differences as far as individual lexical choice is concerned. The findings indicate that general words occur with high frequency in both corpora, but tend to be used to a higher extent in the popular subcorpus. The conclusion that can be drawn from this observation is consistent with the idea that the two genres are interrelated.

4.3.3 Keywords

The majority of studies which make use of wordlists also compare frequency lists more or less manually (e.g. Gledhill 2000). However, there is an increasing number of researchers who do not make use of individual wordlists but of one keyword list with the help of which the corpora are automatically compared in terms of frequency differences (e.g. Berber Sardinha 1999, Scott and Tribble 2005). The keyword function compares all the words of (two) opposing lists and reports those appearing significantly more often in one collection of texts than in the other. This also includes words which appear more than a minimum number of times in one corpus but do not appear in the other at all. This procedure can be very useful to give an indication about the distinctiveness of two corpora, but it can obviously not give any evidence in terms of similarities. Thus, relying on the keyword lists alone is, in my opinion, not a sensible move for tackling the question if and how two genres are related. The word regarded as *key* will point out what is unique but not what is shared and as the pilot study has shown, this can be misleading. In this connection, Scott utters the following words of caution:

> There can be no guarantee that the "key" words are "key" in the sense which you may attach to "key". An "important" word might occur once only in a text. They are merely the words which are outstandingly frequent or infrequent in comparison with the reference corpus. (Scott 1999: WordSmith Manual)

Even if the *keyword* function alone does not suffice for a comprehensive characterisation of the lexical processes between the two corpora, *keywords* can help to identify low frequency words that would otherwise be left aside. Therefore, I would argue in favour of adding prominent low-frequency items to the subsequent qualitative analysis. Even words that occur less frequently can give a good indication about the nature of the texts and, furthermore, words that are less commonly used can pose a particular difficulty to understanding.

By using *Wordsmith*'s keyword function, two lists will be obtained that indicate words that are statistically significant in their frequency of occurrence as compared to the individual reference corpus. Keywords are calculated by using log-likelihood statistic (Dunning 1993). It is assumed to give a better estimate of keyness than the classic chi-square test of significance. To obtain the keyword statistic the following procedures are carried out:

- calculating the frequency for each word in both corpora,
- counting the absolute number of tokens in each corpus,
- matching the two statistics obtained, i.e. word frequency & corpus size,
- cross-tabulating the results and applying the log-likelihood test,
- producing a list of (in our case positive) keywords for each corpus.[15]

As far as the different sizes of two subcorpora are concerned, Tribble (1998) and Xiao/McEnery (2005) explain that the size of a corpus from which the lists are created is not decisive (although the reference corpus is usually larger) and that reliable results can be obtained even if a corpus is bigger than the corpus it is compared with. Tribble also claims that "one feature you notice when comparing a keyword list with a frequency list for the same data is that they are

[15] Scott (1999, *WordSmith* Manual) defines a positive keyword as follows: "a word which is positively key occurs more often than would be expected by chance in comparison with the reference corpus."

usually very different" (Tribble 1998: 7). As the top 100 keywords for both corpora in Table (23) and (24) show, this is not necessarily true of the two collections of texts under investigation:

Table 23: *Keywords for the popular subcorpus*

N	Word	N	Word	N	Word	N	Word
1	PEOPLE	26	DEADLY	51	CAME	76	ALREADY
2	OFFICIALS	27	SAY	52	BRITAIN	77	SCHOOL
3	YESTERDAY	28	SHANGHAI	53	CHIEF	78	MASKS
4	CHINA	29	ORGANISATION	54	PARTY	79	LIU
5	GOVERNMENT	30	OFFICIAL	55	SCHOOLS	80	ASKED
6	SAYS	31	TOLD	56	HEALTH	81	EPIDEMIC
7	TODAY	32	NEWS	57	WARNED	82	HEYMANN
8	DOCTORS	33	COME	58	CAPITAL	83	TRYING
9	LAST	34	HONG	59	SICK	84	CHEN
10	BEIJING	35	KONG	60	AROUND	85	SUFFERING
11	CHINESE	36	KNOW	61	STOHR	86	FEARS
12	FLU	37	GOING	62	MONTH	87	EXPERTS
13	CITY	38	HARD	63	STUDENTS	88	PUT
14	WORLD	39	ANNOUNCED	64	PANIC	89	THOUSANDS
15	SCIENTISTS	40	HIT	65	BELIEVED	90	TOLL
16	DISEASE	41	SPREAD	66	TAKE	91	EAST
17	JUST	42	DIRECTOR	67	AGO	92	CALLED
18	HOW	43	COUNTRY	68	FOREIGN	93	SOUTHERN
19	ASIA	44	YEAR	69	GERBERDING	94	MYSTERIOUS
20	VICTIMS	45	STILL	70	THOUGH	95	MR
21	DOWN	46	WARNING	71	HERE	96	KILLER
22	GET	47	WEEK	72	GETTING	97	WOMAN
23	FAR	48	SPREADING	73	LITTLE	98	AGENCY
24	KILLED	49	UK	74	THINK	99	BRITISH
25	DOCTOR	50	OFF	75	GO	100	BAD

Compared with the results of the refined frequency lists for the top 100 most frequent items, there are words that are regarded as key, such as *disease*, *world*, *patients* or *coronavirus*. They are not only high frequency words but are actually shared by both genres. Nevertheless, as they turn out to be key for one subcorpus, the calculation indicates that they are shared with such different frequencies that it is considered as statistically significant. This is not only a valuable outcome as far as methodology is concerned, but is also helpful for the categorisation of possible lexical processes as it provides a means to ascertain which concepts are more important for a particular genre. There are also some items such as *people* in the popular subcorpus and *COV* in the medical subcorpus that have already been identified as unique in the previous analysis.

Table 34: *Keywords for the medical subcorpus*

N	Word	N	Word	N	Word	N	Word
1	PATIENTS	26	ANTIBODY	51	FEATURES	76	EPIDEMIOLOGIC
2	CLINICAL	27	VIRAL	52	SEQUENCES	77	NUCLEOTIDE
3	PATIENT	28	FIGURE	53	SAMPLES	78	USE
4	ASSOCIATED	29	LABORATORY	54	CRITERIA	79	DISTRESS
5	CORONAVIRUS	30	RADIOGRAPH	55	COHORT	80	G
6	COV	31	INFECTION	56	NASOPHARYNGEAL	81	IDENTIFIED
7	CARE	32	ISOLATES	57	SEQUENCE	82	PRIMERS
8	PERSONS	33	DEFINITION	58	FINDINGS	83	PRESENTATION
9	PROBABLE	34	MG	59	MEDIAN	84	W
10	ONSET	35	EXPOSURE	60	USED	85	ML
11	HCWS	36	RIBAVIRIN	61	NEGATIVE	86	B
12	TRANSMISSION	37	NOVEL	62	AVAILABLE	87	ELEVATED
13	ICU	38	ANALYSIS	63	INCLUDED	88	OBTAINED
14	SERUM	39	MECHANICAL	64	CELL	89	SERA
15	RESPIRATORY	40	RNA	65	H	90	AGENT
16	CASE	41	SARS	66	LYMPHOPENIA	91	SECONDARY
17	PCR	42	DATA	67	CONVALESCENT	92	INTRAVENOUS
18	INDEX	43	TABLE	68	VERO	93	SEROLOGIC
19	ADMISSION	44	SPECIFIC	69	CULTURE	94	MACAQUES
20	CHEST	45	DURING	70	ALVEOLAR	95	INFILTRATES
21	CELLS	46	INTERFERON	71	MAR	96	REQUIRED
22	C	47	RT	72	RADIOGRAPHS	97	MARCH
23	SPECIMENS	48	POSITIVE	73	METAPNEUMO-VIRUS	98	INTUBATION
24	ARDS	49	DAYS	74	INITIAL	99	INVESTIGATIONS
25	VENTILATION	50	CONTACTS	75	OUTCOME	100	DYSPNEA

Now, their keyness status seems to underline their particular role. Nevertheless, the majority of keywords is made up of actual low frequency items and they illustrate the differences one would intuitively expect to find between the popular and the scientific representation of SARS. So it is obvious that the keyword function sheds some empirical light on the intuition about some of the lexical differences among the genres.

The first lexical words that are key in the popular subcorpus and have not been included in the previous top 100 list, i.e. *victims*, *killed* and *deadly*, are those that clearly show the difference in terms of connotations. They may be key indicators more of style than of 'aboutness' as this negative prosody is a consistent feature throughout the keywordlist for the popular texts. Nouns such as *panic* and *fears* and the adjectives *mysterious*, *foreign* or *bad* express a certain, undoubtedly negative bias. As the newspress certainly seeks to place such a bias on solid ground, officials are not only referred to as *scientists* but also as *experts* and they utter words of *warning* which trigger the emotive language that embeds these quotes. To underline this, there is also for the first time a sound group of proper names (*Stohr*, *Gerberding*, *Heymann*) that actually gives the names of the individual experts who are quoted. As far as proper names are concerned, the reference to affected regions is extended and specified, as not only *Asia* is given but also *Shanghai*, and there is a group of words that are related to Britain. This is of course due to the British papers, Guardian and Observer, that contributed texts to the popular subcorpus. All in all, the low frequency keywords are of a general nature and express a clear negative prosody. This is in direct opposition to the keywords in the medical subcorpus which are highly specific terms.

Those words that are unique to the medical corpus in low frequencies are free from connotations as they are not only part of a specialised terminology but are also morphosyntactically complex, such as *nasopharyngeal*, *metapneumovirus*, and they receive their meaning from the component parts. In contrast, there are specific items that are not complex terms but can also act as barriers to comprehension. There are 15 abbreviations in total consisting of acronyms, *HCWS*, *ICU* or *ARDS*, and single letters that are not clitic items but abbreviations that are used autonomously, such as *C* for *Celsius*. Thematically, there is a distinct group of words that actually gives an idea of the different 'aboutness' of the medical texts since there is a great number of terms that not only refer to virus identification but moreover to genetic sequencing (*specimen*, *nucleotide*, *sequences*, *cohort*). Contrary to the popular subcorpus, details on the diagnosis and treatment of patients is given (*radiograph*, *ventilation*, *Ribavirin*, *intravenous*).

All in all, there are striking differences between the two keyword lists that can be best described in terms of dichotomies:

- general, simple words vs. specific, complex terms;

- negative prosodic words versus neutral terminology;
- thematically: report of events (quotes) versus presentation of findings and application (data).

The fact that such words were identified as key makes them promising candidates for detailed concordance-based analysis. The study of their context could also provide an explanation why such words have cropped up with unusual frequencies. Although I would argue that the detailed analysis of the top 100 lists already serves as a valid basis for the following quantitative analysis, I will add prominent low frequency keywords to the categories already established. By doing so, the comprehensive model can take accurate account of the shared and of the individual lexis in the popular and scientific representations of the SARS discourse.

4.4 Summary and classification of quantitative relation(s)

The present study aims to answer two central questions: (1) Is there a process that can be regarded as scientific discourse popularisation? (2) How does the lexis of medical publications relate to it? The quantitative study has shown that the two varieties of the SARS discourse do not exist as separate entities but share a considerable amount of common lexical ground which is tied to the topic and to shared semantic fields.

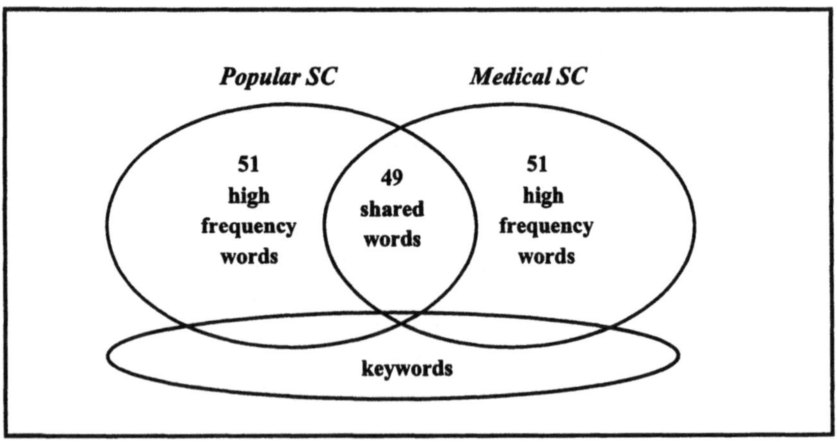

Figure 18: *Quantitative relation(s)*

Although there is a striking consistency in the lexical inventories in the two corpora with regard to the top 100 words, there is, however, also a striking difference in high frequency and low frequency items which reflects the nature of the different genres. Figure (18) illustrates these quantitative relations.

Methodologically, the three different steps (raw frequency, refined frequency and keywords) that lead to these results have proved most effective for a comprehensive view on the two subcorpora. I would thus argue that each of the three steps is obligatory since raw frequency data has to be regarded with some reservation. To make the simple tool of wordlisting most effective, it is essential to employ a stoplist that excludes function words that are also shared by general corpora, such as the FLOB corpus, and edit the lists manually. I would also argue in favour of taking the 100 most frequent items as a basis but combine them with prominent low frequency words. Bringing together the quantitative results so far, the basic categories that were established in the pilot study are still valid, but need to be refined:

Category I: Words that are shared by both corpora

- with similar frequency (*virus*);
- with differing frequency and are thus regarded as key to one genre (*world*).

Category II: Individual words that are unique to one subcorpus

- in high frequency and are therefore partly regarded as key (*people*);
- in low frequency and are entirely regarded as key (*victims*).

Category III: Structurally related words and in both corpora that are realised by

- a different word form of the same lemma (*patient vs. patients*);
- a simplified form of a specific term (*flu vs. influenza*).

The analysis has shown that there is variation within the classification itself since differences in the semantic reference and in the choice of specific terminology vs. general language words exists. As it stands, the quantitative analysis has, despite minor drawbacks, provided a useful bird's-eye view on the nature of the texts comprised by each genre. The general perspective is a valuable basis as it gives information about the significance of individual words and their distribution within the corpora. However, while pure frequency results have proven useful for the categorisation it should not be the only criterion by which the lexical processes in scientific discourse popularisation can be characterised. The questions raised will now be answered in more detail by taking the lexical environment of prominent candidates into account.

5 Qualitative Analysis

The famous quote from Firth (1957: 179) *you shall know a word by the company it keeps* prepared the ground for a contextual theory of meaning which places utterances in their situational context and in their verbal co-text.[16] The lexical unit is influenced by its neighbouring parts and contributes at the same time to their very structure and choice. The fact that item and environment mutually determine each other is referred to as *co-selection* (cf. Sinclair 1991). To picture this interrelation between the two genres, I intend to create a lexical profile that focuses on the lexical unit itself and that expands towards single collocates, possible multiword clusters and larger co-textual chunks that allow for the interpretation of the contextual, co-textual and co-selectional patterns.

The precondition for such a profile is a systematic relation between an item and its co-text. In corpus linguistics, research into lexical profiling is a fairly recent development. Sinclair (1996 and ff.) thinks that each word possesses its own lexicogrammatical profile which may be defined as the sum of information contained in its environment. Hoey (2005) follows this approach and introduces the theory of *lexical priming*, a theory often regarded to offer a new view on words and language. Hoey argues that every word is *primed* to be used in a particular discourse. The priming is a direct result of "cumulative effects of different encounters" with a word and, most importantly, one of the effects is that regular word sequences are constructed that are in turn also primed (Hoey 2005: 13). Every word has a distinct profile of primed sets which consist of collocation, colligation and semantic associations. Therefore, it is evident that the examination of words in isolation cannot provide a reliable description of such processes and, consequently, a comprehensive model of scientific discourse popularisation needs to draw on an extended lexical profile.

The umbrella research question stated in section 2.3 asks to what extent the lexis in the popular presentation of the SARS discourse is related to its medical counterpart. The comparison of both genres in terms of single word frequency has already pointed towards potential links but has at the same time found some fundamental differences. In the contextual study, these initial results will be further refined by a computer-based, corpus-linguistic analysis of concordance lines and collocations for selected words. The resulting lexical characterisation of a term will provide an answer to the questions listed below which are specifications of the questions formulated in Chapter 2. They are structured in accordance with the results obtained in Chapter 4.

---

[16] I will use the term *co-text* to refer to the immediate verbal environment and the term *context* to refer to the situational parameters including genre, register, medium and topic.

Category I: Words in common

- Are shared words used in the same semantic context and with a similar co-text?
- Is there a systematic agreement regarding their collocational profile?
- Do identical forms show an identical semantic prosody?

Category II: Words apart

- Do the individual words belong to individual semantic fields?
- Is there a systematic disagreement regarding their collocational profile?
- Is there a clear-cut difference in semantic prosody; i.e. positive and negative connotations?

Category III: Words related

- Are related words used synonymously?
- Do related words share preferred collocates?
- Is there an agreement or rather a difference in their semantic prosody?

To answer these questions, I will first take a brief look at the basic concepts in lexicology and lexical semantics, in particular, at the syntagmatic and paradigmatic level of lexis. I will then proceed with the more detailed corpus analysis in which lexical profiles for selected words will be set up with the help of concordance fragments and statistically computed collocates and clusters.[17]

5.1 The study of meaning in context

As pointed out in the previous chapter, the focus of the present study is on lexical words and not on grammatical words, because the former are referential and meaning-carrying. It goes without saying that function words carry grammatical meaning and are indispensable for the structure and the success of any form of discourse. However, for the present study it is far more interesting to pinpoint those factors which actually motivate lexical choice in the dual representation of SARS. Which words, for example, are chosen to denominate the causative agent of the disease as *metapneumovirus*, *coronavirus*, *virus*, *bug*, or *killer*? To get a better idea of the processes that lead to the preference of certain lexical structures, a linguistic analysis of the presumed paradigmatic and syntagmatic relationships will be conducted. Before proceeding with the actual lexical profiling, I would like to discuss some basic principles that I will refer to

[17] In order to analyse clusters coherently, I will limit the span to three-word clusters as this is the highest cluster that is yielded in the calculation of the keywords with relative low frequency.

in the course of the analysis. Firstly, the notion of paradigmatic and syntagmatic relationships between words, as illustrated in Figure (19), is of prime importance.

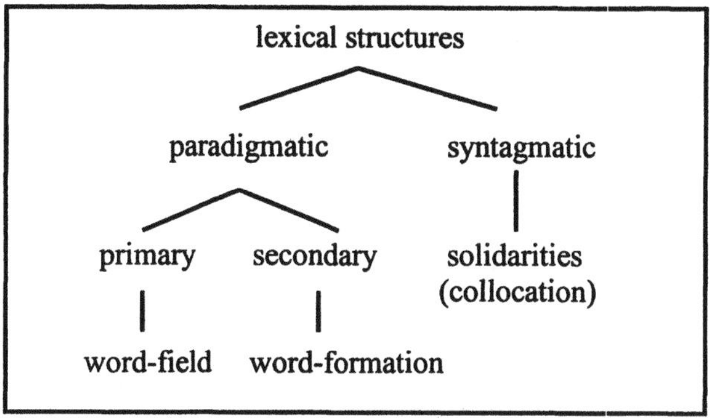

Figure 19: *Lexical structures (adapted from Lipka 1992: 11)*

The structural relationship between words is either of paradigmatic (i.e oppositional) or of syntagmatic (i.e. combinatorial) nature. Paradigmatically, individual lexemes can be compared on a vertical axis. This means that structural relations, such as homonymy, polysemy, synonymy, hyponymy or antonymy can be identified. These relationships may result in a network of co-selective patterns and, as Esser (1999: 165) points out, this network of paradigmatic choices can be quantitatively described by applying corpus-linguistic methods. As stated at the beginning of this chapter, the detection of parallel patterns is one of the main objectives of research into scientific discourse popularisation.

It is often assumed that synonymy plays a significant role in popularisation since scientific terms are thought to be substituted by simpler vernacular words (e.g. substituting *epistaxis* with *nosebleed*). Nonetheless, it is obvious that complete or absolute synonymy is virtually non-existent. According to Lyons (1981: 149), two words can only be regarded as absolutely synonymous "if and only if they have the same distribution and are completely synonymous in all their meanings and in all their contexts of occurrence". Therefore, I prefer to call two items *near synonymous* in order to stress that they denote a similar concept or have a common reference. It is furthermore vital to make a clear distinction between hyponymy and hyperonymy. Partington (1998: 32), for example, gives corpus evidence that it is possible to replace a hyponymous term with its superordinate without altering the truth value of the utterance, whereas the opposite process might indeed alter the truth, e.g. replacing *virus* by *coronavirus*. This question of entailment is of particular interest as it could help in the development of a model of scientific discourse popularisation. As far as other paradigmatic relations are concerned, the question arises whether a systematic relationship of opposition, i.e. antonymy, between words in the two corpora can be established. The quantitative analysis has suggested that there are merely some general, gradable words that can be explicitly or implicitly compared, e.g. *high* and *low* or *many* and *few*. The interpretation of concordances for such scalar pairs might reveal whether they are unmarked in this case (positive and unbiased) or whether they are marked (negative and biased).

The syntagmatic relation between words refers to the horizontal level and is of a combinatorial nature, because it describes solidarities between different words, such as collocations. Paradigmatic and syntagmatic levels are equally important when it comes to *semantic prosody*. This term was first introduced by Louw (1993) and suggests that words can have certain semantic sets or notions with which they frequently occur. The semantic notion often shows a positive or negative polarity as illustrated by Stubbs (1995). He analyses collocations for the word *cause* and demonstrates that more than 90% of the co-occurring words have a negative prosody, e.g. *accident*, *cancer* or *crisis*. Stubbs (1995) and Hoey (1997, 2002), among others, focus on words that collocate with semantic categories or sets. Stubbs (1995) has proven that the word *unemployment*, for example, collocates regularly with words from the field of statistics. In connection to this, Partington (2004) believes that lexical choice from a certain set of items which are available on the paradigmatic scale is most often an unconscious process to express lexical evaluation or stance. For example, he points out that the maximizer *utterly* collocates with unfavourable words that can be brought together under the labels of "absence of quality" (*helpless*, *useless*, *unable*) and "change of state" (*failed*, *ruined*, *destroyed*) (Partington 2004: 147).

So, semantic prosody can be defined in terms of connotations, i.e. in relation to positive or negative prosodies (Louw 1993, Sinclair 1991, Stubbs 1995), and in terms of denotations, i.e. in relation to the preference for lexical or semantic sets (Stubbs 1995, Hoey 1997, Partington 1998). It will be interesting to find out whether this can also be applied to the analysis of lexical processes in the course of scientific discourse popularisation. The concept of semantic prosody, which occurs when a word associates with a particular set of meanings, is of importance but has so far been neglected in any large-scale popularisation analysis. The knowledge of semantic prosody in scientific language has mainly been applied to metaphors. I use the concept of prosody to gain an overview of the lexical relations that the words in the two subcorpora habitually form. For this reason, I use semantic prosody in a wider sense that includes connotations as well as semantic sets.

Returning to the syntagmatic view on lexis, not only does it look at the combinatorial relationship between different words on the horizontal axis, but it furthermore tries to provide a systematic account of the meaning of words which tend to occur together. Concordance lines present a fragment of these horizontal combinations, and collocations can expose regular affiliations statistically. Firth (1957: 196) further expresses his view on collocational meaning, by arguing that it is "an abstraction at the syntagmatic level and that it is not directly concerned with the conceptual or idea approach to the meaning of word". In response, Leech (1974) explains that an actual part of the meaning of a word is that it collocates with other words. In addition, he outlines seven types of meaning of which one is the collocational one. In corpus linguistics, units of meaning have experienced constant redefinition and extension. Sinclair (1987, 1991, 2004) was among the first who broke new ground in using corpus-linguistic evidence for describing phraseology. Two of his key views on the organisation of language can be subsumed under the idiom principle and the open choice principle. The idiom principle states that a language user has a large number of "preconstructed or semi-preconstructed phrases available that constitute single choices, even though they appear to be analysable into segments" (Sinclair 1991: 109). The open choice principle on the other hand sees language as a result of a large number of complex choices. "At each point where a unit is completed (a word or a phrase or a clause), a large range of choices opens up and the only restraint is grammaticalness" (Sinclair 1991: 110). It is the idiom principle that provides an *explicans* for phraseological phenomena that can be genre-specific.

Collocations and concordances can give an indication of lexical choice and recurrent word patterns as they show if a word has systematic association with other words. The simple but effective tool of concordances can also highlight patterns of lexis and word meaning by showing the immediate lexical surroundings of a word.[18] Concordance lines allow retrieval of all of the occurrences of a particular type together with a small proportion of its co-text. Once potentially interesting words are identified, those types can be examined in context which will reveal more about their meaning, function and usage. In addition, an insight into meaning and usage will give an idea which words appear together in the particular genre. The concordance itself is a useful instrument to gather statistical evidence by processing and identifying collocates. Hunston describes collocation as follows:[19]

> Collocation is the tendency of words to be biased in the way they occur. [...]It can be considered as the tendency of two words to co-

[18] Common concordancers are, for example, *WordSmith Tools*, *MonoConc* or *ParaConc*.

[19] For a comprehensive overview of the concept of collocation see, for example, Nesselhauf 2005.

> occur, or as the tendency of one word to attract the other. [...] Collocation may be observed formally in any instance of language, but it is more reliable to measure it statistically, and for this a corpus is essential. (Hunston 2002: 68)

Collocation describes the mutual tendency of habitual co-occurrence of words that is independent of word class or syntactic structure. Collocations can be divided into lexical and grammatical collocations. The latter are referred to as *colligations*. According to Sinclair (1991), colligations are phrases that consist of a dominant word (noun, adjective or verb) and a preposition or grammatical structure, such as an infinitive or clause. Lexical collocations, in contrast to colligations, typically consist of nouns, verbs, adjectives and adverbs.

Hill (1999) extends Hymes' concept of *communicative competence* (as discussed in Chapter 2) by adding the idea of *collocational competence*. He thinks that associations or links between words can be seen as a characteristic of any genre or restricted field of discourse. Knowledge of these verbal routines is thus essential in order to participate in the discourse in an idiomatic way. Bowker and Pearson (2002) claim that idiomaticity is one type of LSP word behaviour that can be easily identified with the help of collocations in a specialised corpus. Nelson (2000) supports this as he gives corpus evidence that certain collocations but also clusters were idiomatic for the LSP variety that he was studying, i.e. economic texts. Clusters are strings of words that contain and are built around the node. They are complementary patterns that do not necessarily carry lexical meaning as they typically consist of a noun phrase (determiner plus noun). Nevertheless, they can give evidence for possible extended units of meaning as they display words that tend to occur in combination. Linguistic research on special languages has already suggested (see Chapter 2) that longer multiword clusters are more frequent and specific to the scientific genre. Therefore, shorter clusters which serve a multi-functional purpose are likely to be expected in the popular texts. As will be shown in section 5.3, the collocations and clusters found in the respective subcorpora give an indication of possible relations between the popular and the scientific representation of the SARS discourse.

5.2 Category I: Words in common

It is neither feasible nor beneficial to discuss all occurrences of the words discussed in Chapter 4 individually. Instead, I will use a practical approach to cope with large amounts of data as suggested, for example, by Sinclair (1999) or Hunston (2002). If too many lines are yielded, *hypothesis testing* allows for taking a small selection that is henceforth used as a reliable basis for a set hypothesis of patterns. First, I will take 30 randomly selected lines and observe detailed patterns, then cross-check these patterns by looking at a larger proportion and select 100 lines for general patterns. As a result, the selected examples can be extracted for further discussion that confirm or contradict the general tendency. To this end, categories were already established at the end of the last chapter which classify the significant words found preliminarily. For the three categories, words are now to be chosen for further qualitative analysis. This choice will be based on statistical reasoning (frequency) and on intuition (prominence).

5.2.1 Shared words with a similar distribution

The frequency analysis of the two subcorpora has revealed a prominent number of words that are shared by both genres but it has also shown that some words in this group differ significantly in their distribution. As a result, the top 100 items have been recognised as keywords by the software. I will first consider those words for a collocational analysis that have a similar ranking or a close relative frequency. Table (25) lists the ten most frequent shared words with a similar distribution in both corpora:

Table 25: *Shared words with a similar frequency*

N	Word	Pop SC Rank (Freq.)	Med SC Rank (Freq.)
1	SARS	1 (1.80%)	1 (1.61%)
2	HEALTH	2 (0.58%)	4 (0.41%)
3	VIRUS	5 (0.44%)	7 (0.32%)
4	CASES	6 (0.41%)	3 (0.57%)
5	HOSPITAL	11 (0.27%)	12 (0.29%)
6	WHO	12b (0.23%)	17b (0.22%)
7	OUTBREAK	16 (0.19%)	19 (0.19%)
8	REPORTED	19 (0.16%)	15 (0.23%)
9	SYMPTOMS	24 (0.14%)	27 (0.16%)
10	CONTROL	37 (0.11%)	30 (0.15%)

The top position in the medical and in the popular subcorpus is held by *SARS*, with the next item following after a wide frequency gap. This does not come as a surprise since *SARS* is not only the agreed on denomination of the disease but also the overall topic marker. At a distance of over one percent, ***health***, ***virus*** and ***cases*** are follow-ups which all have a similar ranking among the top ten lexical words in both corpora. Among them, *virus* stands out as a semi-technical term, meaning that it originates from the medical domain but is institutionalised in general language use. *SARS* and *virus* are chosen for the following detailed analysis.

5.2.1.1 SARS

At the beginning of the outbreak, the cause of the illness was unknown and the disease was referred to under shifting names. The most common medical denomination was *atypical pneumonia* and the general news press used the name of the familiar *flu* or called it a *flu-like*, or *pneumonia-like illness*. In March 2003, the WHO designated the disease as *Severe Acute Respiratory Syndrome* (SARS). The acronym was almost immediately used independently and without its accompanying complete denomination. This is also evident in the corpus as SARS is yielded as the most frequent item but the individual parts of the name rank much lower in frequency. It is generally used in form of an initialism, but as the following example shows, it is also used as a full-fledged proper name with capitalised first letter.

(1) What is Severe Acute Respiratory Syndrome (Sars)? It is the name being used to describe the serious disease with pneumonia-like symptoms that first emerged in southern China, [...]. <Guardian April 2>

The Guardian used this word-formation from the beginning and consistently for the entire period of their SARS coverage. Note that in the medical subcorpus SARS is only used as an initialism and was never institutionalised the way it was done in the general press. I will now take a detailed look at the lexical environment of SARS. Figure (20) gives ten concordance lines with *SARS* as the node for the popular subcorpus:

N	Concordance		
1	acute respiratory syndrome	*(SARS)*	. A flurry of new cases in
2	of the first doctor to identify	*SARS*	. Rumors that the virus is
3	from every region hit by	*SARS*	may severely overestimate
4	are a reaction to the spread of	*SARS*	outside of Thailand
5	ending on Sunday, confirmed	*SARS*	cases in the Amoy Gardens
6	finally to admit that the	*SARS*	virus has spread through
7	non-medical people infected by	*SARS*	quietly rocketed. The
8	say over the weekend that	*SARS*	may still be an epidemic
9	illness widely believed to be	*SARS*	affected Guangdong province
10	statistics. According to WHO,	*SARS*	has also spread inland

Figure 20: *Concordance sample for* SARS *in the popular subcorpus*

The fragment already shows that the node word *SARS* is part of identifiable patterns. It is not only used as an abbreviation added to the full denomination (line 1), but it is for the most part used separately as an autonomous term. When looking to the left and to the right of the node, patterns become apparent. SARS is active in *hitting* (line 3) and *affecting* regions (line 9). It *spreads* both *inland* and *outside* of single countries (lines 4, 6, 10) and it *infects* people, which leads to new *cases* (lines 7, 1, 5). It is moreover used to modify, or more exactly, to specify the virus and the cases by combining words into expressions such as *SARS virus* (lines 5, 6). Statistically, *SARS* co-occurs regularly with the lexical words given in Table (36). I have once more removed grammatical function words (*the*, *of*, *to*, *in*, *a*, *and*) from the list. By omitting these, the resulting collocates focus not only on the syntagmatic relation between collocation and node but also draw attention to their paradigmatic relation.

Table 26: *Collocates of* SARS *in the popular subcorpus*[20]

N	Word	Total	Total Left	Total Right
1	CASES	313	94	219
2	VIRUS	233	46	187
3	PATIENTS	204	24	180
4	WHO	196	118	78
5	RESPIRATORY	180	134	46
6	SEVERE	172	126	46
7	SYNDROME	171	130	41
8	ACUTE	169	125	44
9	HEALTH	156	68	88
10	OUTBREAK	151	39	112

The quantitative data in Table (26) confirm the observation that the abbreviation is already institutionalised as the agreed denomination and is used to great extent independently of its component parts. *Severe, acute respiratory* and *syndrome* are common collocates (ranks 5, 6, 7, 8) but are not yielded as most frequent co-occurring words. Considering that SARS was a relatively brief epidemic – from March to October 2003 – the productivity in a relatively short period of time draws attention to the significance of acronyms for word-formation processes in scientific discourse popularisation. *Cases* is the most frequently occurring lexical collocate of *SARS* in the popular subcorpus and appears 313 times in

[20] I have adopted *WordSmith*'s preset standard span of calculation which is within five words to the left and to the right of the node.

total. As example (2) shows, it is used to refer to people that have been infected by the virus.

(2) It is unclear how many of those are true SARS cases because there has been no accurate way to test for the infection. <Washington Post April 14_2>

In referring to people as *cases*, a neutral stance and a rather impersonal position is expressed. This is in line with the pilot study, which has already suggested that generalisation is a significant, yet genre-untypical feature (cf. the example of *patients* in Chapter 3.4). The average reader can usually relate to an abstract event much more easily if it is personalised and, to make abstract events more appealing to the general public, the newspress tends to rely on the strategy of personalisation. It is certainly of interest for the research question posed to investigate in more detail whether the feature of generalisation in the popular subcorpus is an actual outcome of the popularisation process.

The third most frequent collocate of SARS is *patients*. The word is used in a similar way as *cases* in that it also refers to affected persons:

(3) The WHO then estimated the capital's SARS patients at 100 to 200—five times more than the government acknowledged—based on hospital visits and other data. <Time Asia April 28_5>

Contrary to *cases, patients* is used in connection with numerical data and especially with high figures. The example indicates why *WHO* is one of the most frequent collocates of *SARS*. In the majority of occurrences, the presentation of figures in relation to *SARS* is associated with officials or institutions, such as the *WHO*. The initialism of *World Health Organisation* plays a key role in the popular coverage of the disease as it is constantly presented and cited as a reliable source for detailed information about the epidemic, as the following example demonstrates:

(4) The world appears to be winning the war against Sars, the World Health Organisation (WHO) announced yesterday, though China's outbreak is still cause for major concern. <Guardian April 29>

This might account for the frequency of *health* in connection with *SARS*. In the vast majority of occurrences, *health* either is a component of the organisation's name or refers to health officials, in particular, to spokespersons of the *WHO*. The short form itself is, very much like *SARS*, used independently in the corpus without the component parts. Example (4) furthermore reveals that *SARS* associates with words that have a clearly negative prosody. The containment of the disease is pictured as warfare[21] and the outbreak of SARS causes major concern. Example (5), which gives the co-text of the frequent collocate *spread*, supports this view, too:

(5) Even so, he warned that dilapidated medical facilities, poor equipment and shoddy monitoring for epidemics were hidden dangers for the spread of SARS in the countryside. <Time Asia May 19>

Again an official is quoted with words of warning. In this way, fear is fuelled that deficient infection control techniques in the rural parts of China might pose a threat to global public health. The transmission of SARS was actually halted within three months but it was the initial uncertainty whether the virus could be identified and controlled that dominated the coverage in the early stage. The silent transmission reinforced this tendency as example (6) illustrates:

(6) It is possible that testing could reveal some people have been infected with the SARS virus without becoming very sick. <SMH April 12>

Reflecting its significance for the popular discourse, *virus* is the second most important collocate of SARS and is repeatedly used to refer to the causative agent of the disease in a general way.

The study of SARS as a node-word in the popular corpus has shown that it regularly associates with particular meanings or, more accurately, with semantic sets. It is used as an accompanying acronym which is added to the denomination and it is used as an agent which affects people and actively spreads. In terms of connotations, there is an apparent negative prosody as SARS is associated with uncertainty and danger that causes concern and worry. Since the comparison of the wordlists of the two corpora has revealed an agreement in the exceptional distribution of SARS, it must now be investigated whether this frequency-based similarity also holds true for the co-text. The following concordance is a fragment of randomly selected concordance lines for SARS in the medical subcorpus. In comparison with the popular subcorpus, there are some apparently identical but also some divergent patterns. Once more, the acronym SARS serves as a supplement to the full name of the disease (line 1), and affected persons are again referred to as *cases* or *patients* (lines 2, 3). The causal agent is identified as the *SARS virus* but also, more specifically, as *coronavirus* (lines 4, 8). Apart from using the hyponym rather than the superordinate term *virus*, there are other features that are not as prominent in the popular subcorpus.

---

[21] The 'war metaphor' is common in the coverage of infectious diseases. I will comment on this in more detail in section 5.3.

N	Concordance		
1	acute respiratory syndrome	*(SARS)*	shows no sign of abating
2	There were 60 patients with	*SARS*	and there had been seven
3	three reported cases of	*SARS*	on a daily basis and
4	that the causal agent of	*SARS*	is a new coronavirus
5	or suspected to be linked to	*SARS*	were in Ontario. In
6	trauma centres, closed its	*SARS*	and critical care units
7	in Beijing to investigate.	*SARS*	is still present in China
8	about the origin of the	*SARS*	virus and the likely
9	the standard treatment for	*SARS*	to a combination of
10	great need to strengthen	*SARS*	control and surveillance

Figure 21*: Concordance sample for* SARS *in the medical subcorpus*

The medical counterpart further raises issues of the *treatment* of the disease (line 9), which is in part carried out in special SARS *units* (line 6). However, it also discusses the need for *control and surveillance* of the disease (line10). Table (27) provides an overview of the lexical collocates of *SARS* in the medical subcorpus.

Table 27: *Collocates of* SARS *in the medical subcorpus*

N	Word	Total	Total Left	Total Right
1	PATIENTS	514	291	223
2	COV	365	11	354
3	CASES	345	146	199
4	PROBABLE	263	239	24
5	WHO	221	118	103
6	CORONAVIRUS	215	70	145
7	RESPIRATORY	200	174	26
8	SEVERE	188	161	27
9	SYNDROME	181	173	8
10	ACUTE	180	170	10

As expected, the most evident pattern parallel to the collocates of *SARS* in the popular subcorpus is the co-occurrence of the component parts *severe*, *acute*, *respiratory* and *syndrome*. The higher rank of *respiratory* is a result of its role as an adjective in the medical texts which means that it can either refer to the *respiratory tract* or to *respiratory symptoms*. Although *WHO* collocates with *SARS* in both corpora to the same extent, its evenly split distribution, 118 times to the left versus 103 times to the right, suggests a difference. Example (7) shows *patients*, the most frequent collocate for SARS in the medical corpus, in combination with the collocate *WHO* which can either refer to the organisation or to the relative pronoun:

(7) WHO case definitions for suspected SARS have a negative predictive value of 85% and a sensitivity of 26% for detecting SARS in patients who have not been admitted to hospital <BMJ April 21_2>

Patients also appears in connection with numerical information which is usually given succinctly in the form of relative percentages. What is striking is the similarity in which *cases*, as shown in example (8), is used in both corpora without any specific figures. *Cases* occurs frequently in connection with another *SARS* collocate, namely with the adjective *probable*:

(8) GPs have been advised that if a patient fits the definition of suspected or probable cases of SARS, the patient should be seen at home if at all possible, rather than in the practice setting. <BMJ May 3>

Probable and *possible* are neutral markers of uncertainty and, all in all, the co-text does not contain items with an obvious negative semantic prosody.

The second most frequent collocate, the abbreviation *CoV* (for *coronavirus*; also *COV*) occurs 354 times to the right of the node and only eleven times to its left, which indicates that it is predominantly used as part of the compound *SARS-CoV*, as shown in example (9). I have already pointed out that hyphenation is used ambiguously in the different journals as each journal has its own printing convention. The Journal of the American Medical Association, for example, never fuses the two items to one hyphenated term, but always uses them as unconnected terms. *CoV* is the common abbreviation for *coronavirus*, the specific type of virus that causes the disease and *coronavirus* itself also collocates with *SARS*, but to a lesser extent.

(9) The identification of the SARS-associated coronavirus (SARS-CoV) has led to the development of serologic and virologic tests for the disease. <CMAJ June 10>

(10) The worldwide outbreak of severe acute respiratory syndrome (SARS) is associated with a newly discovered coronavirus. <Lancet July 27>

It does not come as a surprise that in connection with SARS the medical texts prefer the subordinate and thus more specific term *coronavirus* over the more general term *virus*. Researchers have been working hard on the identification of the previously unknown variant and the specific causative agent is of major importance to the diagnosis, treatment and containment of the disease.

Regarding the lexical profile of SARS, the study of its contextual features in both corpora has revealed a general agreement in so far as the acronym regularly associates with particular sets of words with similar denotations. *SARS* combines with them to an adjunct to the name cluster (*Severe Acute Respiratory Syndrome*) to specify the agent of the disease (*SARS virus* or *CoV*) and to refer to affected people (*SARS cases* or *patients*). As Table (28), the calculation of three-word clusters for SARS provides further evidence of this similarity.

Table 28: *Clusters for* SARS

	Popular SC	Medical SC
1	ACUTE RESPIRATORY SYNDROME	PATIENTS WITH SARS
2	THE SARS VIRUS	ACUTE RESPIRATORY SYNDROME
3	CASES OF SARS	CASES OF SARS
4	THE SARS OUTBREAK	SARS ASSOCIATED CORONAVIRUS
5	SPREAD OF SARS	TRANSMISSION OF SARS

Nevertheless, some differences (or rather genre specific preferences) can be observed in the lexical choice. Two out of the five cluster examples in Table (28) denote the same concept but either use general words (*SARS virus, spread of SARS*) or more specific or technical terms (*SARS associated coronavirus*; *transmission of SARS*). Although the lexical profiles for SARS in the two collections of texts seem to be similar in terms of common semantic sets, they fundamentally differ in their connotations, given that the examples taken from the popular subcorpus show a distinct negative semantic prosody. Words like *war*, *warning* and *danger* have not been used in the corresponding context in the medical subcorpus at all.

SARS is the most common shared word and has unquestionably an outstanding role as it is not only the agreed on denomination of the disease, but also functions as a topic marker for the discourse in both genres. It is now interesting to see how other shared words like *virus* behave.

5.2.1.2 VIRUS

The quantitative analysis suggests that the two subcorpora apparently share a considerable amount of general language items, but also some topic-related words that originate from the medical domain. *Virus* is a good example of such a term, although it has been institutionalised in everyday-language use and is therefore regarded as semi- or sub-technical (cf. Busch 1994, Fischer 1998). In Chapter 2, it was pointed out that one of the core traits of technical or scientific terms is that they possess a clear-cut referential meaning and are unambiguous and context- independent. Whether this holds true for the semi-technical term *virus* or whether it acquires a new idiomatic function in the popular texts has to be tested.

N	Concordance		
1	identify SARS, rumors that the	*virus*	is airborne, and fears
2	finally to admit that the SARS	*virus*	has spread throughout the
3	residents not to worry as the	*virus*	was spreading only within
4	not told they might carry the	*virus*	now Henan has more than
5	the likely source of the	*virus*	and the worst hit.
6	SARS is believed to be a	*virus*	, in their quest for new
7	they cannot filter away any	*virus*	and therefore cannot protect
8	warned of the severity of the	*virus*	and the possibility that
9	if he had arranged for the	*virus*	to hold off on its assault
10	is generally believed that the	*virus*	itself has not been

Figure 22: *Concordance sample for* VIRUS *in the popular subcorpus*

We have already seen that *SARS* and *virus* frequently collocate in the popular texts, but the fragment shows that although the specific SARS virus is meant, the term is recurrently used autonomously. On the left of the node, different associations and attributes become apparent. The virus is something *severe* which has to be *warned* about (line 8) and about which *rumors* spread (line 1). Furthermore, tentative language, i.e. epistemic modals, are used to express uncertainty: the virus is *believed* to be the *likely* source (lines 6, 10, 5) and it *might* be carried (line 4). On the right side of the node, the virus takes action as it *spreads*, *hits* and *is airborne* (lines 2, 3, 1).

Table 29: *Collocates of* VIRUS *in the popular subcorpus*

N	Word	Total	Total Left	Total Right
1	SARS	230	184	46
2	SPREAD	86	37	49
3	NEW	60	46	14
4	COULD	56	8	48
5	MORE	54	13	41
6	PEOPLE	46	23	23
7	DEADLY	44	38	6
8	INFECTED	42	25	17
9	MAY	42	11	31
10	ABOUT	40	34	6

The lexical collocates provided in Table (29) for *virus* confirm this first impression. In accordance with the previous results, SARS is the most frequent collocate for *virus* and occurs an astounding 230 times in total. In the majority of cases, they turn up jointly in the form of the already familiar two-word cluster *SARS virus*. The second most frequent collocate is *spread* and, as the following examples illustrate, not only the virus itself spreads, but also stories and rumours *about* it.

(11) The under-reporting of the spread of the virus is being widely denounced, with little attempt to deny there had been deliberate concealment. <Guardian April 22>

(12) Rumours about the virus have spread on the internet and through mobile text messages. <Guardian April 25_6>

The two passages stress that this is due to the suppression of official information about the disease which was an (unfortunate) deliberate policy in the early days of the epidemic, when the virus was still *new*.

(13) As this strange new virus continues its spree, killing hundreds and infecting thousands more, scientists are working overtime, trying to keep people from harm. <Newsweek May 5>

Example (13) illustrates that the fairly neutral collocate *new* is embedded in a highly negative environment. It is not only *new*, but moreover pictured as a strange killer which harms large numbers of people and keeps scientists busy in protecting the public. This negative prosody is also reflected in the frequent co-occurrence of *deadly*.

(14) China has been unable to wipe out the deadly new virus. <Washington Post May 22>

(15) China, has in many ways held on to the values that Singapore abandoned to face the deadly virus head-on. <Asia Times April 8>

This collocate dominates in discussions of the severity, scale and management of the disease. As already outlined, the deadly virus is often said to be met with ineffective measures. Whenever it is attributed as *deadly*, the fierce nature and the severe impact of the virus are stressed. Severity seems to play an important role, which is also stressed when we look at two further frequent collocates, *people* and *infected*, which tend to co-occur.

(16) So far, the virus has infected 2,890 people. <Washington Post April 14>

Cases usually occurs in connection with hardly any figures, *patients* with rising numbers and now *people* seems to be used with global figures. Linking this kind of statistical benchmark with high numbers and the general expression *people* makes the impact of the virus more relevant to the reader than it may actually be. Moreover, the numbers are given out of their immediate context, without specifying e.g. the source, the area and time of spread. This seems to support the notion of vagueness that is repeatedly employed. The epistemic modality that was already observed in the concordance fragments is confirmed by the frequent co-occurrence of virus with the modals *may* and *could*.

(17) In Hong Kong, doctors believe the virus may have mutated when it infected Amoy Gardens residents, who suffer unusual symptoms (including severe diarrhoea). <Time Asia May 5_4>

(18) The big concern in this area of the world is that one day another influenza virus could hop the barrier between animals and humans. <SMH March 19>

In examples (17) and (18) the co-selection of modals signals the level of uncertainty about the information that is given. The quantitative analysis in Chapter 4 has shown that the use and distribution of modals (*may*, *might*, *can*, *could*, *would*) is largely restricted to the popular representation of SARS. Out of this group only *may* and *might* are also used in the medical texts, but with lower frequency. According to Palmer (1990: 5) the function of epistemic modals is to assess the possibility concerning the truth value of a proposition. This explains the predominance of such modals in the public discussion of the epidemic since their use leads to the expression of assumptions rather than facts. By doing so, the newspress can variously speculate about the source and the risks of the new virus. This reflects not only the hypothetical nature of the statements made but can also be regarded as a strategy to maintain caution by avoiding the assertion of unjustifiable statements (as in example 17). This is in line with Hyland's observation (1998) who claims that the main rhetorical functions of epistemic modals is to hedge 'evidence' in order to signal that judgements made by the author should be regarded as tentative only. Yet, it is the very tentativeness that allows the readers' imagination to interpret the information in a negative way. In addition, a clear negative semantic prosody is reflected by the use of expressions like *suffer* or *big concern*.

Apart from the negative prosody, the general agreement in the semantic sets for SARS in both genres suggests that the shared frequency could be an indication for some shared lexicosemantic features. This will now be put to the test by matching the distinctive popular patterns for *virus* with its distribution in the medical subcorpus. Consider the case of *virus* as given in Figure (23).

N	Concordance		
1	of infection due to the SARS	*virus*	and are therefore good markers
2	clinical response to the	*virus*	in a high contact environment
3	genome of two strains of the	*virus*	believed to be responsible
4	CoV, the largest of any RNA	*virus*	. It is possible but unlikely
5	have shown that the SARS	*virus*	is mutating at a rapid rate.
6	laboratory conditions the	*virus*	can survive in urine for at
7	possible hosts for the SARS	*virus*	include bears, monkeys, and
8	doesn't understand how the	*virus*	is being spread and has not
9	the genetic map of the SARS	*virus*	say that although it is
10	any known coronaviruses, the	*virus*	is unlikely to have its origins

Figure 23: *Concordance sample for* VIRUS *in the medical subcorpus*

When we look to the right of the node, it becomes apparent that in half of the concordance lines the type of agent is specified by pre-modifying *virus* with *SARS* or *RNA*. In addition, the explicit variants *coronavirus* or *CoV* supplement the general term (lines 10, 4). In contrast to the popular subcorpus, *virus* appears in connection with its genetic properties and the *genome of two strains* is held responsible (line 3) which is believed to have *mutated* (line 5) and it is therefore crucial to identify the *genetic map* of the virus (line 9). Consequently, the virus is tested under *laboratory conditions* and a *clinical response* can be triggered (lines 6, 10). In a similar vein to the popular subcorpus, an indication for vagueness and uncertainty can be noted, since the virus is *believed* to be responsible (line 3) and *possible* hosts as well as *unlikely* sources are given (line 7, 10).

Table 30: *Collocates of* VIRUS *in the medical subcorpus*

N	Word	Total	Total Left	Total Right
1	SARS	120	95	25
2	RESPIRATORY	51	40	11
3	CORONAVIRUS	46	24	22
4	SYNCYTIAL	35	32	3
5	INFECTION	33	12	21
6	HEPATITIS	30	24	6
7	HUMAN	30	14	16
8	ISOLATION	29	3	26
9	PORCINE	26	16	10
10	PCR	24	15	9

There is an indication for both, a similar and a different lexical distribution of the term *virus* in the two genres, yet, the collocates of *virus* in the medical subcorpus, provided in Table (30), point in a different direction. Surprisingly, the only shared collocate for *virus* in both subcorpora is *SARS*. The remaining collocates are specific to the medical domain. *SARS* however, is used as frequently as in the popular texts as part of the compound *SARS-virus*. This is in accordance with the concordance samples already shown. The second most frequent collocate *respiratory* has to be treated ambiguously, since it can either refer to the respiratory disease itself, to the *respiratory tract* (example 21) or to the specific kind of *respiratory virus* (example 22) that causes SARS:

(19) The finding of infectious virus in the respiratory tract supports this contention. <Lancet April 19_3>

(20) Apparently it has a broad spectrum of activity on RNA viruses and has been approved for the treatment of respiratory syncytial virus and hepatitis C in conjunction with interferon. <CMAJ June 24>

Example 22 also gives an indication why the terms *syncytial* and *hepatitis* are frequent collocates of *virus* in the medical texts.[22] In taxonomy, *coronavirus* is the subtype of a virus, named after its crown-like shape. The coronavirus itself has several individual varieties, one of which is the *respiratory syncytial virus* and the *hepatitis virus*. This group of co-hyponyms can be further extended by other collocates yielded in the analysis. *Human* can be categorized in the same line with *porcine* (from pigs) and *murine* (from mice). As the following examples show, all of them refer to specific viruses that belong to the same family.

(21) The virus belongs to antigenic group 2 of the coronaviruses, along with murine hepatitis virus and bovine coronavirus. <Lancet April 19_3>

(22) The three major coronavirus antigenic groups (I, II, and III), represented by [...] porcine epidemic diarrhoea virus (PEDV), human coronavirus OC43 (HCoV-OC43), bovine coronavirus (BCoV) [...]. <NEJM May 15>

Whereas *virus* associates with related semantic sets in the medical texts in order to narrow and refine its denotation, the popular realisation remains at a higher level of abstraction and makes use of hyperonyms. In this connection, it is also obvious why *isolates* and *PCR* (*polymerase chain reaction*) collocate frequently with *virus*. The medical texts present the crucial steps towards the identification

---

[22] *syncytial* = property of certain viruses to form cellular bundles.

of the causative agent under laboratory conditions, as illustrated in example (23).

(23) Samples from lung, duodenum, jejunum, kidney, liver, and spleen were collected post mortem for virus isolation and RT-PCR. <Lancet July 27>

First, the virus variant needs to be isolated from samples obtained and then PCR tests can be performed to establish a genetic map. All of this is of major importance for the target audience of the medical texts but it is obviously not relevant to the general public.[23]

When the different realisations of *virus* in the two subcorpora are brought together, hardly any correspondence in terms of their collocational profile or their semantic prosody can be found. As Table 41 shows, the clusters in the medical subcorpus are complex noun phrases (e.g. *porcine epidemic diarrhoea virus*) that embed the general concept of *virus* in a highly specific co-text to arrive at a clear-cut referential meaning and avoid ambiguity.

Table 31: *Clusters for* VIRUS

N	Popular SC	Medical SC
1	THE SARS VIRUS	THE SARS VIRUS
2	VIRUS THAT CAUSES	RESPIRATORY SYNCYTIAL VIRUS
3	SPREAD OF THE	MOUSE HEPATITIS VIRUS
4	THE VIRUS COULD	PORCINE EPIDEMIC DIARRHEA
5	FROM THE VIRUS	EPIDEMIC DIARRHEA VIRUS

Apart from the clear difference that the medical texts display in terms of restricted function, the most obvious distinction between the realisations of *virus* in the various genres can be made in terms of its semantic prosody. Whereas the word in the medical corpus relates to highly specific semantic sets (as given in Table 31), the popular subcorpus does not indicate recurrent semantic sets, but shows prosody in terms of negative connotations. Considering the agreement of the lexical profile of *SARS* and the disagreement of the profiles of *virus* in the popular and in the medical discourse, I presume that popularisation is not about realisations of single terms in isolation but about their contextualisation or, more precisely, their recontextualisation.

[23] PCR tests, or more exactly, RT-PCR tests (*reverse-transcriptase polymerase chain reaction*) are done to confirm directly the presence of coronavirus RNA (*ribonucleic acid*).

5.2.2 Shared words with a different distribution

Identically distributed words can behave fundamentally differently in the two subcorpora. For this reason, it is decisive to find out if keyness can reflect such a notion on a statistic level. The quantitative analysis in chapter 4 has shown that some highly frequent words are shared by both genres, but with a significantly different distribution. The keyword test has supported this quantitatively and has drawn attention to the fact that this might be an indication of a popularisation process. Table (32) ranks the top five high frequency keywords.

Table 32: *Shared high frequency words with a different distribution*

Keyword Popular Subcorpus				Keyword Medical Subcorpus			
N	Word	Rank (Freq.) Pop SC / Med SC		N	Word Med	Rank (Freq.) Pop SC / Med SC	
1	WORLD	12a (0.23%)	85 (0.08%)	1	PATIENTS	13 (0.22%)	2 (0.94%)
3	SPREAD	14 (0.22%)	54 (0.10%)	2	RESPIRATORY	23 (0.14%)	5 (0.36%)
2	DISEASE	3 (0.51%)	12 (0.28%)	3	CORONAVIRUS	99 (0.06%)	9 (0.29%)
4	CHINA	7 (0.40%)	33 (0.13%)	4	CARE	80 (0.08%)	11 (0.29%)
5	INFECTED	15 (0.20%)	39 (0.12%)	5	CASE	44 (0.10%)	8 (0.29%)

Even though the words are shared by both collections of texts, those with a significantly higher frequency in the popular corpus (*world*, *disease*, *spread*, *China*, *infected*) play a greater role for the public representation of SARS. Remarkably, all these words are interrelated, because they all refer to the global scope of the outbreak. In November 2002, the disease was first discovered in the southern parts of China and spread to more than 30 countries by July 2003. The virus infected almost 10,000 people and was officially recognised as a global threat by the WHO. This explains the importance of the top five keywords in the medical subcorpus (*patients*, *respiratory*, *coronavirus*, *care*, *case*), since institutions were responsible to provide case definitions and treatment regimens for the care of patients. I will first look at *world*, as a word that is seen as key for the popular subcorpus and then discuss the case of *coronavirus*, which is key for the medical subcorpus and related to the analysis already, carried out for its superordinate term *virus* that is shared by both genres similarly.

5.2.2.1 World

As in the case of other epidemics (e.g. Ebola or bird-flu), news coverage grows fairly extensively as soon as the global nature of a disease is recognised. The outbreak affected several countries on more than one continent, because SARS spread around the world via air travel. Consequently, it does not come as a surprise that *world* is the most significant high frequency keyword in the popular subcorpus. However, it has to be pointed out that it is a general term and cannot provide us with straightforward insights into the popularisation process, unless it is used in a specific context. The concordance lines provided in Figure (24) show that this might in fact be the case.

N	Concordance		
1	to get back to normal as the	*World*	Health Organisation
2	gets going in the developing	*world*	and there are hundreds
3	occurring anywhere in the	*world*	becomes a concern
4	to David L. Heymann of the	*World*	Health Organization
5	US sneezes, the rest of the	*world*	catches a cold. The
6	will be a good test for the	*World*	Health Organisation's
7	carried by animals around the	*world*	and among the least
8	yet the inspectors of the	*World*	Health Organisation
9	major transmission around the	*world*	. The question is now
10	definition laid out by the	*World*	Health Organisation

Figure 24: *Concordance sample for* WORLD *in the popular subcorpus*

The sample illustrates that *world* is most often used as a part of the name of the *World Health Organisation* (lines 1, 4, 6, 8, 10). If this is not the case, it refers to the global nature of the disease which occurs *around the world,* (lines 7, 9) in the *developing world,* in the *rest of the world* and virtually *anywhere in the world.* Contrary to its general dictionary meaning, *world* is used specifically as part of the name components of the *WHO* or it is used idiomatically in fixed phrases. The collocates listed in Table (33) support this initial observation.

The by far dominating collocation is, of course, *World Health Organisation* (WHO). The following examples are just representatives of the vast majority of those found in the corpus which rely heavily on the WHO as a source of information.

(24) The disease, described by the World Health Organisation (WHO) as 'a serious global threat', has claimed at least 17 lives in the last six weeks. <Observer March 30>

(25) The Chinese government downplayed the problem until it had spread to Hong Kong and beyond, prompting an alert from the World Health Organization in mid-March. <Newsweek April 14>

Table 33: *Collocates of* WORLD *in the popular subcorpus*[24]

N	Word	Total	Total Left	Total Right
1	HEALTH	315	19	296
2	ORGANISATION	140	1	139
3	ORGANIZATION	127	0	127
4	AROUND	105	104	1
5	WHO	85	12	73
6	SARS	63	38	25
7	CHINA	29	12	17
8	DISEASE	29	22	7
9	OFFICIALS	28	12	16
10	CASES	21	14	7

One explanation for the significant role of the WHO in the popular representation of SARS is that it is the advisory and coordinating institution whenever a global epidemic occurs. As already seen, WHO *officials* were also quoted repeatedly in connection with the words of warning when the disease appeared to turn into a global epidemic in March 2002. This was picked up over and over again by the news press in accordance with the principle "bad news sells best". The most frequent collocate of *world* after World Health Organisation is *around*, which is almost exclusively used in the fixed phrasal expression in example (27).

(26) There, he is believed to have infected several people who then began the widespread transmission of SARS around the world. <Time Asia April 21>

As the following lines show, the remaining collocates *disease*, *China* and *cases* are used in similar contexts as they relate not to the global scope but more to the heavy impact of the disease on China.

(27) At least 2,305 people in China have been infected with SARS and 106 have died, the largest incidence of the disease in the world. <Washington Post April 24_2>

(28) Additionally, countries such as China, with the largest number of cases in the world. <Washington Post April 25>

[24] Spelling is treated differently in the popular texts. The British press uses British English spelling (*organisation*) and the US prefers American spelling (*organization*).

All in all, the use of *world* seems fairly restricted and it can be described best in terms of idiomaticity than in terms of semantic prosody. It is key for the popular subcorpus for two reasons: first, and most importantly, because it belongs to the larger cluster *World Health Organisation* which is of major importance as a source of information for the popular press; second, it is used to refer to the global dimension of the disease which transformed into an epidemic posing a threat to people all around the globe.

The previous analyses of SARS and *virus* have shown that a seemingly identically distributed word can have a similar or a different lexical profile in the two genres. Since *world* functions in a special way in the popular subcorpus, it is interesting to find out whether this is also true for the scientific subcorpus. The concordance sample provided in Figure (25) indicates that the use might indeed be comparable.

N	Concordance		
1	to official figures. The	*World*	Health Organization has
2	China's fall in SARS cases. The	*World*	Health Organization's
3	of 13 laboratories around the	*world*	. The pace of SARS
4	have recovered. On May 23 the	*World*	Health Organization
5	third worst affected area in the	*world*	. More than 241 new
6	by using criteria from the	*World*	Health Organization and
7	days before symptom onset, to a	*World*	Health Organization (WHO)
8	have generally followed the	*World*	Health Organization (WHO)
9	of SARS in a globalized	*world*	also includes several
10	and clinicians around the	*world*	have worked collaboratively

Figure 25: *Concordance sample for* WORLD *in the medical subcorpus*

It is most obvious that *world* is used in both genres in a common way, mainly as part of the name of the *World Health Organisation* (lines 1, 2, 4, 6, 7, 8). Furthermore, it apparently co-occurs with *China* and *cases* (line 2) and is used idiomatically in the already familiar phrasal expression *around the world* (lines 3, 10). When we look to the left of the node, some new features are observable. Words that are common in medical discourse – *laboratory*, *criteria*, *symptom onset* and *clinicians* – have not been used in connection with *world* in the popular texts. The collocates listed in Table (34) prove that although *world* is regarded as key in one of the two subcorpora, its usage and function seem to be almost equal. As expected, the most frequent collocation is that of *World Health Organization*. In contrast to the popular texts, the medical corpus shows a consistent spelling, i.e. the American English variety. The reason for that can be found in the fact that American spelling is considered as the norm in scientific articles that are published in magazines with an international scope, such as the journals included in the corpus.[25]

Table 34: *Collocates of* WORLD *in the medical subcorpus*

N	Word	Total	Total Left	Total Right
1	HEALTH	83	12	71
2	ORGANIZATION	62	0	62
3	SARS	21	11	10
4	OTHER	18	6	12
5	AROUND	17	17	0
6	DEPARTMENTS	9	9	0
7	LOCAL	9	9	0
8	CONTINUING	7	1	6
9	CASES	6	2	4
10	REPORTED	6	4	2

Examples (29) and (30) underline the importance of *WHO* also for the medical representation of SARS.

(29) A total of 8,427 probable SARS cases were reported to the World Health Organization (WHO) from 29 countries. <MMWR July 18>

(30) CDC continues to work with state and local health departments, the World Health Organization (WHO), and other partners to investigate cases of severe acute respiratory syndrome (SARS). <MMWR July 4>

As is typical of epidemic outbreaks, scientific information and data related to the disease change almost daily. Researchers and practitioners respond to the worldwide outbreak and the WHO gathers and circulates relevant information like the number of infected persons. The collocates *cases*, *reported* and *local health departments* should therefore be interpreted in close connection to this. The organisation is equally important to the public and to the scientific audience of the medical journals. Nonetheless, for the latter, it can be not only the source of information but also a partner (example 30) in the collaborative effort to stop the epidemic. An additional correspondence with the popular corpus is the reference to the global scope of the disease. The collocates *around* and *other* relate to affected areas or parts of the world apart from China:

25 The spelling norm is codified in general publication guidelines, provided by publishers such as *Elsevier* and databases such as *PubMed* or *MEDLINE,* and is also given in the style sheets of the respective journals.

(31) SARS CoV isolates from Guangzhou shared the same origin with those in other countries, and had a phylogenetic pathway that matched the spread of SARS to the other parts of the world. <Lancet Oct 28<

(32) Laboratories around the world are continuing to collaborate in the global effort to identify this causative agent. <WER March 21>

Once more, the two corpora not only share the same semantic associations but also show a similar idiomaticity. Although *world* is regarded as key, and is thus more significant for the public representation of SARS from a quantitive perspective, it is used almost identically in both genres. The 3-word clusters in Table (35) provide further statistical evidence for this. For this reason, keyness should not be seen as an indication for differences in two (collections of) texts as it obviously just reports on the special importance of a word for one genre.

Table 35: *Clusters for* WORLD

N	Popular SC	Medical SC
1	WORLD HEALTH ORGANISATION / WORLD HEALTH ORGANIZATION	WORLD HEALTH ORGANIZATION
2	AROUND THE WORLD	HEALTH ORGANIZATION WHO
3	HEALTH ORGANISATION WHO / HEALTH ORGANIZATION WHO	AROUND THE WORLD
4	REST OF THE	DEPARTMENTS THE WORLD
5	PARTS OF THE	AND LOCAL HEALTH

Despite all the similarities that have been identified so far, we have to consider that *world* is a general term and the question arises whether such words behave as general terms in both genres. The previous analysis has suggested that semi-technical terms which derive from the medical domain (like *virus*) are used differently in the popular and in the scientific representation of SARS. It is now necessary to find out how a term is issued that is ranked at a lower level of abstraction (like *coronavirus*) and that is regarded as key in the medical subcorpus.

5.2.2.2 Coronavirus

Coronavirus is a medical term that, unlike its hyperonym *virus*, is not common in everyday language use. The study of *virus* has shown that it differs to a large extent in its lexical profiles, even though the word is used similarly frequently in the two corpora. Here, *coronavirus* ranks in both wordlists among the top 100 items, yet with such a differing frequency (rank 9 in the medical subcorpus vs. rank 99 in the popular subcorpus) that it is clearly regarded as key for the scientific genre. The investigation of *world* has shown, however, that keyness is not necessarily a qualitative feature but rather a quantitative one. It is thus interesting to see if the technical term behaves more like its hyperonym or rather more like the general co-keyword *world*. As pointed out in chapter 2.2, Busch has put the idea forward that terms can be *determinologised* (Busch 1994: passim) by losing at least some of their original semantic scope when they are transferred to general language. Tognini Bonelli also supports the view that this can happen virtually to any word when it is *delexicalised* (Tognini Bonelli 2002: 116). As the concordance lines in Figure (26) show, *coronavirus* seems to co-occur with at least some related semi-technical terms in the popular subcorpus:

N	Concordance		
1	SARS is a member of the	*coronavirus*	family, a new pathogen
2	year - provided that the new	*coronavirus*	doesn't turn out to
3	caused by a microbe known as	*coronavirus*	. Such viruses usually
4	known family of viruses: the	*coronavirus*	, the probable cause
5	nasty version of a feline	*coronavirus*	kills every cat it
6	but researchers said the	*coronavirus*	increasingly appeared
7	by a new variant of the	*coronavirus*	similar to the common
8	the leading suspect is a	*coronavirus*	that jumped from cows
9	even more worried that the	*coronavirus*	which may be mutated
10	particular family of viruses:	*coronavirus*	. It was the corona

Figure 26: *Concordance sample for* CORONAVIRUS *in the popular subcorpus*

The sample already provides several essential attributes of the node. Most importantly, the coronavirus belongs to a *known family of viruses* (lines 1, 4, 10), although it is a *new variant* (lines 2, 7), or, more biased, a *nasty* feline *version* (line 5). There are some fairly neutral descriptions that make use of medical terminology, such as *probable cause*, *new pathogen* or a *microbe* (lines 1, 3, 4) which has *mutated* (line 9), but there are also negative prosodies. The coronavirus is pictured as a *leading suspect* that *kills* and causes *worry* (lines 5, 8, 9). At first sight, the lexical profile is two-fold: on one the one hand, there is the embedding of the term in an explanatory context, on the other hand, there is the rather negative association when general words are employed. Yet, the collocations provided in Table (36) show that the explanatory co-text seems to prevail.

The concordance lines suggest that the most frequent collocate SARS is again used as a noun compound with *coronavirus* as shown in example (33), but also, as in example (34), in connection with the frequent collocate *family*.

Table 36: *Collocates of* CORONAVIRUS *in the popular subcorpus*

N	Word	Total	Total Left	Total Right
1	SARS	53	23	30
2	NEW	25	23	2
3	CAUSES	18	0	18
4	FAMILY	16	3	13
5	CAUSE	15	2	13
6	CAUSED	14	12	2
7	FOUND	12	9	3
8	KNOWN	11	6	5
9	MORE	10	3	7
10	SCIENTISTS	10	4	6

(33) The SARS coronavirus is the 14th known member of a family of viruses named for their distinctive, crown-like shape. <Time Asia June 2_2>

(34) One is from the coronavirus family of viruses, the other from a different family called paramyxoviruses. <Washington Post May 29>

The examples illustrate that the term is embedded in a sequence of text that is neither predominantly technical nor general, but balanced in order to introduce the new concept to the reader. This is a pattern that can also be observed when the importance of *coronavirus* as the likely source of SARS is discussed, which is reflected by the frequent collocation of the node with the individual word-forms of *cause*:

(35) Currently, the favourite theory is that it is caused by a coronavirus, often the cause of the common cold, which can spread quickly but for which there is no cure. <Guardian April 2>

(36) Scientists now believe they have found the likely culprit--a previously unknown member of the coronavirus family, which also causes colds. <Newsweek April 7>

(37) This week scientists in several centres reported that SARS is most likely caused by a virulent new coronavirus, the type of virus that causes influenza, and laboratory researchers around the world raced to break its genetic code. <SMH April 14>

Apart from the explanatory context, there are at least two notable additional features. First, the coronavirus is often introduced with the help of the other more common illness that this type of virus is held responsible for, namely *influenza*, generally referred to in non-medical settings as *flu* or *the (common) cold.* I assume that this will be different in the medical subcorpus, because it is a known fact which is not required for facilitating comprehension in the same way as it definitely does in the popular subcorpus. The other recurrent feature is the quoting of scientists or experts when the concept of coronavirus is discussed. Only *scientists* is yielded as a frequent collocate, but, as the following examples show, *experts* is used in a similar way:

(38) Experts found coronavirus in feces and on an infected resident's toilet. <Newsweek May 5>

(39) SARS is caused by a previously unknown coronavirus that scientists suspect jumped from an animal to humans in southern China in November. <Washington Post May 9>

The examples further indicate that *coronavirus*, in contrast to its hyperonym *virus*, shows no recurrent negative prosody but rather an association to semantic sets that facilitate comprehension. These sets seem to reflect the neutrality of the medical source domain and this particular process of embedding might account for the fact that the term is *negatively key* (that is, less frequent) in the popular subcorpus. The examples prove that *coronavirus* is actually a foregrounded item, which is marked as unfamiliar and is hardly ever used without its explanatory context. If this is done, it is exclusively for cohesive purposes and negative prosody is shown by some general terms that are employed (e.g. *leading suspect*). This is assumed to be different in the medical texts corpus, because *coronavirus* is a familiar concept for scientists. The term is regarded as a keyword in the medical subcorpus and is indeed used more independently, as the concordance sample in Figure (27) shows.

N	**Concordance**		
1	speed which with this novel	*coronavirus*	was detected, character
2	outbreaks associated with	*coronavirus*	is that we cannot
3	and treatment of other	*coronavirus*	diseases of animals
4	diagnosis of SARS-associated	*coronavirus*	infection in most
5	Characterization of the novel	*coronavirus*	associated with SARS
6	in the transcription of	*coronavirus*	RNA with high
7	laboratory evidence of	*coronavirus*	infection into the
8	human metapneumovirus or a	*coronavirus*	may be possible cause
9	test for SARS-associated	*coronavirus*	, effective infection
10	to detect this novel	*coronavirus*	antigen, we can use

Figure 27: *Concordance sample for* CORONAVIRUS *in the medical subcorpus*

First of all, *coronavirus* is not straightforwardly mentioned in connection with its family, yet it is also characterised as *novel* (lines 1, 5, 10), which is an interesting parallel to the use of *new* in the popular subcorpus. If it co-occurs with *SARS* it is also used repeatedly with *associated* (lines 4, 5, 9). Semantically, coronavirus appears in a moderately specific setting with reference to *laboratory evidence* (line 7) and *transcription* of the virus *RNA* (line 6), or is associated with a similar type, the *human metapneumovirus* (line 8). The collocations for *coronavirus* are, with the exception of SARS and *new*, different from those of the popular subcorpus.

Table 37: *Collocates of* CORONAVIRUS *in the medical subcorpus*

N	Word	Total	Total Left	Total Right
1	SARS	214	143	71
2	ASSOCIATED	98	84	14
3	NOVEL	88	86	2
4	INFECTION	57	24	33
5	PATIENTS	51	10	41
6	HUMAN	46	36	10
7	VIRUS	46	22	24
8	COV	36	2	34
9	NEW	33	30	3
10	ISOLATED	31	7	24

Again, the most frequent collocate for the node word is SARS, which modifies *coronavirus* as it enters into a more or less fixed meaningful expression with two other collocates: *associated* and, to a lesser extent, the acronym *CoV*:

(40) [...] concurrent illnesses that were either confused with disease caused by SARS-associated coronavirus or that coinfected patients. <JAMA June 4_2>

(41) Until now, immunohistochemical detection of SARS-associated coronavirus (SARS-CoV) in SARS-associated lesions has not succeeded. <Lancet July 27>

Apart from the highly specific context, the obvious morphosyntactic complexity (e.g. *immunohistochemical detection of SARS-associated coronavirus*) accounts for the particularity of the medical representation of SARS. The preference for the use of *coronavirus*, which is indicated by its keyword role, reflects the norms of a community which expects certain structures and preferred lexical choices expressed in such rhetorical devices. This is also mirrored by the preference for the adjective *novel* to its equivalent *new*. As far as the seemingly synonymous collocates are concerned, the corpus gives evidence that they are in fact truly synonymous as they are used in almost identical settings with a close identical co-text, although *novel* seems to attract medical terms rather than general ones. The following examples illustrate this relationship:

(42) A novel coronavirus has been identified as the aetiological agent of the syndrome. <BMJ June 21>

(43) A new coronavirus has been identified in patients with severe acute respiratory syndrome (SARS). <Lancet June 14_2>

It is a striking fact that the medical texts have a clear preference for the Latinised variety, whereas the popular texts only use the common form of Germanic origin. This supports one of the most common, but at the same time most questioned popularisation strategies which says that in the process classical terms are avoided, or transformed into familiar forms (see Chapter 2.3).

Despite all differences, there is some room for similarities. Example (44) shows that the next two collocates, *infection* and *patients*, display identical patterns in the two subcorpora:

(44) Of the 16 index patients, one (6%) tested positive for coronavirus infection, three (19%) have had a negative serologic test on a convalescent specimen; results from 12 (75%) are pending. <NEJM May 15>

Here, the concept of affected persons is realised by *patients* that have a *coronavirus infection* and, again, this is backed up by providing numerical information. Similar to the popular realisation, *patients* associates neither with vague quantifiers nor with high numbers, but with moderate figures.

Moreover, there is an interesting process within the medical subcorpus itself. The two collocates *human* (example 45) and *isolated* (example 46) seem to bridge the gap between the otherwise different usage patterns of the hyponym *coronavirus* and its hyperonym *virus*.

(45) [...], porcine epidemic diarrhea virus (PEDV), human coronavirus OC43 (HCoV-OC43), bovine coronavirus (BCoV), porcine hemagglutinating encephalomyelitis virus (HEV), [...] <NEJM May 15>

(46) Based on the first full-length genome sequence of a coronavirus isolated from a SARS patient in Canada. <Lancet Oct 28>

Both words (*virus* and *coronavirus*) appear in a similar context and serve a similar function. The superordinate term *virus* is also used in clusters and *coronavirus* joins with *bovine, human* and *OC43* into a complex noun phrase that embeds the concept in a highly specific co-text. As an outcome, even the acronym CoV is used in an extended form as it is merged with the abbreviations of the component parts to form *HCoV-OC43*.

All in all, the medical term coronavirus cannot be regarded as 'delexicalised' but rather as 'recontextualised', because it is introduced to the reader by being used in conjunction with words that have a semi-technical rather than a general notion. The clusters provided in Table (38) also show an agreement in terms of regular associations with words in the direct vicinity of the node:

Table 38: *Clusters for* CORONAVIRUS

N	Popular subcorpus	Medical subcorpus
1	THE CORONAVIRUS FAMILY	SARS ASSOCIATED CORONAVIRUS
2	CAUSED BY A	A NOVEL CORONAVIRUS
3	THE SARS CORONAVIRUS	CORONAVIRUS SARS COV
4	THE NEW CORONAVIRUS	THE NEW CORONAVIRUS
5	MEMBER OF THE	INFECTION WITH THE

Despite all similarities, there is a clear-cut tendency to broaden the meaning in the general context by embedding it in an explanatory context without specifying it any further (e.g. *member of the coronavirus family*), whereas, in the medical texts, there is a tendency to specify and restrict the reference by giving the exact co-hyponyms and denominations (e.g. *bovine coronavirus* and *human coronavirus OC43*).

The findings support the view that a keyword list may prove more useful than a simple wordlist, because it only yields words that occur more frequently in one corpus and can thus underline the fundamental nature of a particular genre. However, this holds only partly true for the words examined since different forms can share important aspects of their profiles.

5.3 Category II: Words apart

Although the research questions asked about the nature of the relation between the two genres, the quantitative study has produced evidence that a lexical popularisation model should not only take account of similarities in both genres, but should to some extent consider what is different. So far, the collocational analysis has shown that similar forms can show differences in their profile and that different forms can display a surprising agreement in their lexical affiliations and semantic associations. The motivation for the investigation of selected words of Category II is two-fold. It is of interest to get an idea of independent word meaning of an item that is chosen for the expression of individual concepts, but it is likewise challenging to test if candidates categorised as "words apart" are interrelated.

5.3.1 High-frequency words

When studying individual words, it is indispensable to take their frequency into account. Earlier corpus-based studies of lexical features have focused on high-frequency words, such as the 100 most recurrent words in a corpus, since they are regarded as particularly relevant to the texts under investigation. Nevertheless, the keyword analysis in chapter 4 has shown that lower frequency items can be prominent features of a discourse. For example, the medical keywords tend to be highly specific and can pose a particular difficulty for a lay audience. I will therefore consider both high frequency words that are particular to one corpus and low frequency keywords. Table (39) ranks the top five words that are among the top 100 words of one corpus only.

Table 39: *Individual high-frequency words*

Popular Subcorpus			**Medical Subcorpus**		
N	Word	Rank (Freq.)	N	Word	Rank (Freq.)
1	PEOPLE	4 (0.46%)	**1**	CLINICAL	18 (0.20%)
2	OFFICIALS	10 (0.28%)	**2**	PROBABLE	20 (0.19%)
3	BEIJING	17 (0.19%)	**3**	TRANSMISSION	22 (0.18%)
4	GOVERNMENT	18 (0.18%)	**4**	COV	24 (0.17%)
5	CHINESE	20 (0.16%)	**5**	ASSOCIATED	31 (0.15%)

The popular subcorpus indicates a surprising consistency as the most frequent individual words refer either to *people* in general or, more specifically to *officials* or collective body of *government* which could be related to *Chinese* and *Beijing*. The medical subcorpus gives a more diverse picture. Three out of the five words are adjectives (*clinical*, *probable*, *associated*) that are particular to the medical texts. With *CoV* being the acronym for *coronavirus*, *transmission* is the only noun proper and we have already seen that it is used as a more technical synonym for *spread* in the popular texts. *People*, as the most frequent general item, and *clinical*, as the most frequent corpus-specific scientific expression are now selected for detailed discussion and lexical profiling.

5.3.1.1 PEOPLE

The brief discussion of *people* in the pilot study has raised some interesting, albeit preliminary, issues. Contrary to one its dictionary meanings, the word does not refer to the population at large but, in a more narrow sense, only to those persons affected by SARS, realised by using the nearly synonymous words *patients* or *people.* A prominent feature of both seems to be the use of vague quantifiers like *many* or *most* which are not used in the medical texts. Furthermore, numbers are only included if they are significant to the popular audience, for example, spectacularly high numbers, such as *1,700 people* who *have been quarantined* (see Chapter 3.4). The concordance sample shows that the earlier findings hold true not only for the fraction of texts used in the pilot study, but also for the popular subcorpus in its entirety.

N	Concordance		
1	least partial isolation of	*people*	suspected of having SARS
2	has killed more than 900	*people*	worldwide since it
3	actor. It's new. It's hurting	*people*	. The question has
4	the virus to as many more	*people*	as possible, allowing
5	there may be large numbers of	*people*	who become infected but
6	underestimated how long	*people*	can carry the virus
7	to be especially lethal to	*people*	60 and older, who could
8	who infects a dozen or more	*people*	shedding the SARS virus
9	(or not yet ill) the infected	*people*	have been around. This
10	talkative. About seven of his	*people*	are working on the virus

Figure 28: *Concordance sample for* PEOPLE *in the popular subcorpus*

The concept of affected persons also stands for people that have been infected by the virus (lines 5, 8, 9) or are at least suspected of having *SARS* (line 1). The only exception to this is found in line (10), where the node refers to people working on the virus like, for example, scientists and laboratory staff. Once they have been infected, *people* can also change their role, linguistically speaking, from patient to agent as they can actively *carry* and *shed* the virus (line 6). When looking to the left of the node, it becomes apparent what can happen to the people as *SARS* not only infects, but it furthermore *hurts* and *kills* them as it is attributed as *lethal* (lines 2, 3, 7). What is again a remarkable feature are the figures and quantifiers that are employed to suggest that SARS is a serious threat to many. The disease affects *large numbers* (line 15) that can be *a dozen or more* or, even fuzzier, *many more* (lines 8, 4) and, in the end, it has killed *more than 900 people* (line 2). These observations are supported by the collocates in Table (40).

Table 40: *Collocates of* PEOPLE *in the popular subcorpus*

N	Word	Total	Total Left	Total Right
1	INFECTED	186	72	114
2	MORE	178	136	42
3	SARS	147	51	96
4	KILLED	107	82	25
5	DIED	90	3	87
6	LEAST	72	56	16
7	SOME	60	52	8
8	MANY	57	48	9
9	WORLDWIDE	57	5	52
10	DISEASE	52	21	31

With 186 co-occurrences, the most common collocate of *people* is *infected.* Interestingly, examples (47) and (48) indicate that they are not only pictured as sufferers of the disease, but they are also presented as a potential risk for public health and a kind of stigmatisation seems to emerge early in the outbreak.

(47) Yesterday, U.S. health officials warned against any unnecessary travel and outlined stringent new measures to keep infected people from spreading the sometimes fatal disease. <Washington Post March 30>

(48) He also believes it is possible that testing could reveal some people have been infected with the SARS virus without becoming very sick. <SMH April 12>

Officials were quoted with words like *warning* and *possible* that signify an underlying anxiety about a disease with an unknown cause and possibly fatal outcome. Identifying potential cases is of course important for controlling transmission and containment, but in the manner in which is was done publicly, fear was not only directed towards the disease itself but also to persons infected with SARS. As far as the disease is concerned, the following examples show that the collocates *SARS*, *killed*, *died* and *worldwide* are used in connection to its risk for the global community.

(49) SARS has killed more than 900 people worldwide since it emerged in China last November. <Washington Post Sep 10>

(50) At least 23 people have died, including two healthy young men with no existing medical conditions. <Guardian May 5>

From its beginning SARS was identified and responded to as a threat to global health. The popular texts present SARS as a major danger and allow it to be faced with (a by and large pointless) panic. As the examples (49) and (50) illustrate, the concept of affected people is extended to those who have been not only infected and have fallen ill, but furthermore to those who finally died.

The second most prominent group of collocates is that of the quantifiers *more*, *many* and *some*, demonstrating not only the vague nature of the utterance but also a possible strategy to employ assertions of uncertain scientific claims. In this connection, we have already seen that epistemic modality is employed to express vagueness (cf. section 5.2.1.2) which is a characteristically device in the public discussion of SARS.

(51) While the disease has killed more than 800 people worldwide and infected more than 8400, the WHO has lifted travel warnings for most SARS-hit areas. <SMH June 20>

(52) The disease is very worrying, and is ruining the lives even of many people who are not infected. <Guardian April 22_2>

Examples (51) and (52) show that use of the words *many* and *more* makes the threat appear larger. Linking this kind of quantifiers with global figures, as in example (51), gives them greater importance than they may in fact have, because the figures are presented out of context. Although the spread of SARS was successfully stopped within a short period of time and the travel warning was lifted (51), in the beginning, it was unclear whether the disease could be kept under control and this threat was constantly highlighted in the popular texts. In example (52), *many* as a vague quantifier allows the readers' interpretation to increase the threat. The clusters for *people* in the popular texts in Table (41) support this view.

Table 41: *Clusters for* PEOPLE *in the popular subcorpus*

N	Cluster	Freq.
1	PEOPLE HAVE DIED	46
2	NUMBER OF PEOPLE	27
3	PEOPLE WHO HAD	24
4	THOUSANDS OF PEOPLE	21
5	KILLED AT LEAST	16

Example (53) shows that the three-word cluster *thousands of people* exaggerates and sets the stage for a potential global jeopardy:

(53) This makes it quite plausible that SARS could kill tens or hundreds of thousands of people in China alone. <Asia Times April 24>

Even though the death rate was under five percent (a fact that was hardly ever stated), there was a possibility that it could become more virulent, a slight risk that it might turn into a pandemic that kills *hundreds and thousands*. The previous examples have shown that this risk has been highlighted repeatedly.

In terms of semantic prosody, *people* associates with certain semantic sets and although the sets themselves do not have a negative connotation (since they are neutral facts), they are used to prompt a negative interpretation of the facts by the audience and therefore evoke negative prosodies. *People* serves as a marker of the vagueness and the unpredictability of the threat in the popular texts as it stresses the concept that nearly everybody could be affected. The reasons for this kind are transparent as in this context the number of affected people can be regarded and presented as newsworthy. This explains why *people* is seen as key for the popular subcorpus but plays no role in the medical domain.

5.3.1.2 CLINICAL

The adjective *clinical* is the most frequent keyword in the medical subcorpus. Although it is quite common in general language use, the results of the previous analyses suggest that it could be used differently as it might diverge from its general dictionary meaning since it does not refer to an institution, i.e. to a clinic or to a hospital. The concordance sample given in Figure (29) confirms this hypothesis.

N	Concordance		
1	A clear picture of its	*clinical*	presentation will help
2	updated epidemiological and	*clinical*	data, as well as
3	important to characterize the	*clinical*	manifestations of
4	result in biological and	*clinical*	differences remains to
5	epidemic. Taken together,	*clinical*	and in-vivo studies
6	of stereoids did not improve	*clinical*	symptoms and signs.
7	have no data on the full	*clinical*	spectrum of the emerging
8	through social contact. Our	*clinical*	description pertains
9	Who are experiencing rapid	*clinical*	progression with severe
10	Where, on the aetiology and	*clinical*	and epidemiological

Figure 29: *Concordance sample for* CLINICAL *in the medical subcorpus*

The concordance lines show that in the medical context, *clinical* is predominantly used as a description of a particular status. When we look to the right of the node, it becomes apparent that it premodifies a specific kind of *data*, *studies* or *spectrum* (lines 1,5, 7) and it also classifies a *description* or *presentation* (lines 1, 8) which is based on observable *manifestations* and *symptoms* (lines 3, 6). When we look to the right of the node, the adjectives that co-occur with the node, *epidemiological* and *biological* (lines 2, 4), give an indication that these potential co-hyponyms are, together with *clinical*, used to refer to certain features that are important for the identification of the virus and for the diagnosis and classification of SARS cases. The collocates provided in Table (42) are in line with this.

Table 42: *Collocates of* CLINICAL *in the medical subcorpus*

N	Word	Total	Total Left	Total Right
1	SARS	77	35	42
2	PATIENTS	60	18	42
3	FEATURES	50	2	48
4	LABORATORY	37	6	31
5	COURSE	24	1	23
6	DATA	23	7	16
7	SAMPLES	22	2	20
8	DISEASE	21	6	15
9	SPECIMENS	21	0	21
10	EPIDEMIOLOGICAL	19	6	13

The collocates for the adjective *clinical* have at least two things in common. With the exception of SARS, they occur most frequently to the left of the node, which maintains its role as a multiple modifier. They are either set medical terms, such as *specimens*, or are general words that are used specifically, i.e. *criteria*. As already outlined, general language items can experience a change or a restriction in their reference when they are used in a scientific context if they are modified by specific adjectives such as *clinical*.

If *clinical* co-occurs with *features* or *criteria*, it is used to refer to the symptoms that an infected patient has. Such clinical symptoms can match not only the case description, but they can furthermore classify him or her as a SARS case and thus actually transfer the status from *probable* to *confirmed*:

(54) Initial reports on severe acute respiratory syndrome (SARS) described the clinical features of confirmed cases. <BMJ June 21_2>

(55) One hospital employee (case 13) had no identifiable direct contact with either patient, but was working at the hospital during the period of unprotected exposure and met the clinical criteria for probable SARS. <CMAJ May 25>

The next group of frequent collocates, *patients*, *disease* and *course*, should be interpreted jointly, as they occur in more or less fixed phrases and refer not only to the symptoms that are brought about by SARS, but moreover to its general development:

(56) Examination of the clinical course of our patients and the frequencies of complications suggests that ribavirin may not have been effective and may actually have caused some of the complications. <CMAJ June 24>

(57) Additional priorities identified included standardization of diagnostic assays and reagents, [...], and development of animal models to improve understanding of pathogenesis and the clinical course of disease. <WER Aug 24>

The clinical course of patients (56) includes their medical history, such as mode and time of transmission, their symptoms and their reaction to treatment such as the administration of the antiviral drug *ribavirin*. The clinical course of a disease (57) involves similar features as the development also includes its source, etiologic agent and effects. The etiologic agent is a key issue for the medical texts and, therefore, not only the *clinical* but more specifically the *laboratory* characteristics are of importance:

(58) Consistent with a novel coronavirus as the etiologic agent, the illness shares clinical and laboratory characteristics with many other respiratory infections. <JAMA May 6>

(59) Laboratory investigations of clinical and postmortem samples from SARS patients, as presented here and in earlier studies (2,5,7,11) already fulfilled the first three criteria. <Lancet July 27>

Example (59) shows that even *samples* can be specified as *clinical*, which has to be seen in opposition to *postmortem*, since the clinical samples have been collected from patients while they were (still) alive. The references (e.g. *earlier studies* or *criteria* in ex. 59) that are given whenever the adjective is employed are quite specific and discuss methods and findings that are highly relevant to the medical audience but not to the general public. The prominence of *clinical* in terms of frequency can therefore be explained in terms of its function and in terms of its semantic associations. As Table (43) shows, the adjective can be combined, alone or alongside with others, into different conceptual units expressed in nominal groups.

Table 43: *Clusters for* CLINICAL *in the medical subcorpus*

N	Cluster	Freq.
1	THE CLINICAL FEATURES	16
2	CLINICAL AND LABORATORY	14
3	CLINICAL AND EPIDEMIOLOGICAL	10
4	THE CLINICAL PRESENTATION	9
5	CLINICAL SPECIMENS FROM	9

So, *clinical* can be used to modify specific or general nouns because it possesses certain inherent qualities that enable the task, namely the combination, extension and linking of those nouns to be performed. Furthermore, the outcome of this combinatorial process refers to issues that are associated with the pathologic identification, treatment and development of SARS and are not discussed in the popular representation of the disease. The actual outcome of clinical studies and the conclusions drawn from clinical results that lead to the containment of the disease are important, but not the methodology used to arrive at this point.

5.3.2 Low-frequency words

Up to now, I have only chosen words for detailed analysis that rank among the top 100 lexical words in the individual lists. But what about words that have a lower frequency but are also promising candidates for lexical profiling? As it is neither feasible nor desirable to assess all items from an extensive list of words, the keyword function of *WordSmith* provides a statistically sound method of additionally taking significant low-frequency items into account.

Table 44: *Individual low-frequency words*

Popular Subcorpus				Medical Subcorpus			
N	Word	Rank (Freq.)	Keyness	N	Word	Rank (Freq.)	Keyness
1	VICTIMS	156 (0.05%)	138.86	1	HCWS	109 (0.07%)	251.14
2	GET	125 (0.06%)	133.24	2	SPECIMENS	110 (0.07%)	197.66
3	KILLED	134 (0.06%)	125.96	3	ARDS	165 (0.05%)	193.61
4	DOCTOR	101 (0.06%)	121.35	4	ANTIBODY	122 (0.06%)	187.60
5	DEADLY	130 (0.06%)	118.98	5	RADIOGRAPH	359 (0.03%)	172.52

Table (44) gives the five most important keywords with a frequency below 100 for each genre. After what we have seen so far, the keywords in the popular subcorpus are actually those one would expect to be given for the popular text, because *victims*, *killed* and *deadly* have a distinct negative prosody that is absent in the medical texts. The keywords in the medical subcorpus seem to have a neutral stance, but this is of course due to their specific nature. All of the five items are restricted to the specific domain and could pose a potential barrier to communication if they are passed on to the general audience. This holds especially true for the prominently positioned acronyms *HCWS* and *ARDS*. The two statistically most distinct items (*victims* and *HCWS*) are now selected for detailed analysis.

5.3.2.1 VICTIMS

The top keyword in the popular subcorpus, *victims*, is of course part of a broader metaphorical frame that depicts the management of the disease as warfare. For a long time, linguists have been interested in exploring the metaphorical framing of illnesses and, since the emergence of AIDS in the 1980s, they have shown a growing interest in the metaphors of infectious diseases and epidemics.[26] In scientific discourse popularisation, a possible language barrier be bridged by using metaphors as they can be employed systematically to "understand one thing in terms of another" (Lakoff and Johnson 1980: 5). Thus, a concept can be expressed by the use of the literal expression or by the use of a metaphor. According to Lakoff and Johnson, and also Lee (2001), they are cognitive models that provide a means for communication partners to create and share understanding. Metaphors are vivid, explicit and easily understandable as they relate the known to the unknown and don't require thorough subject-specific knowledge. The concordance lines in Figure (30) show, however, that this cognitive consideration is not only about *understanding* but even more so about *interpreting* a specific concept in terms of another.

The sample provides a first impression of the framework in which the metaphor is employed. SARS victims *fight* for their lives but are *killed* by the disease (lines 3, 4, 7, 8). They belong to a *group* of victims (line 6) and their *number increases* since *many*, or more specifically *one in ten* victims are killed (lines 1, 4, 7, 8). The stringent countermeasure to keep them from *spreading* the virus (line 2) is the construction of *isolation camps* for up to 1000 victims (lines 5, 9). In this context, it is reasonable that *killed* is the third most frequent keyword in the popular corpus. One of the most important outcome of the study of metaphors is that they tend to occur in groups that form complete metaphorical systems or schemes (Lakoff and Johnson 1980).

---

[26] Biere and Liebert (1997) deal more explicitly with the metaphorical framing of AIDS.

N	Concordance		
1	gained traction and the number of	*victims*	had begun to increase
2	so-called superinfectors - single	*victims*	who spread the disease
3	in a Singapore hospitals, the	*victims*	fight for their lives
4	apparently kills one in 10 of its	*victims*	, more than twice the
5	like an isolation camp for Sars	*victims*	. Taiwan yesterday
6	patients, joining a group of	*victims*	whose deaths have raised
7	diseases which kill many of their	*victims*	quickly are by definition
8	which killed up to 10 of its	*victims*	. The WHO warned that the
9	1000-bed isolation camp for Sars	*victims*	on Beijing's northern
10	time to express sorrow for the	*victims*	, not to celebrate, and

Figure 30: *Concordance sample for* VICTIMS *for popular subcorpus*

The 'war' metaphor with all its negative connotations seems to be a particularly dominant scheme in the public discussion of the nature, actions and impact of the SARS. This is in line with Rijn van Tongeren's (1997) study on metaphors in medical settings. She analysed the kind of metaphors that are used to structure medical concepts and the functions that these expressions fulfil. Based on a corpus of oncological texts, Rijn van Tongeren mapped a coherent group of conceptual metaphors that centre around the concept *of war*: tumour cells *invade*, tumour cells *colonize* and tumour cells are *enemies*. Although the collocates of *victims* provided in Table (45) do not show an explicit negative prosody, their contextual embedding (examples 60-61) supports this metaphorical scheme.

Table 45: *Collocates for* VICTIMS *in the popular subcorpus*

N	Word	Total	Total Left	Total Right
1	SARS	40	39	1
2	MANY	11	8	3
3	ONE	11	9	2
4	WHO	11	2	9
5	SOME	10	8	2
6	HONG	9	4	5
7	KONG	9	5	4
8	MORE	8	2	6
9	NEW	8	4	4
10	LEAST	7	6	1

The most frequent collocate of *victims* is once more *SARS*. Yet, the dominant war metaphor attributes an outstanding active role to the disease and less to the persons affected. This pattern of culpability differs from the emphasis on the individual responsibility apparent in the blame and stigma linked with the high frequency keyword *people*. Example (60) shows that SARS is pictured as a free agent responsible for its actions and those infected are blameless victims that fight against the killer virus.

(60) RACE TO BEAT THE BUG <headline> In a high-security London laboratory, scientists work around the clock to understand the killer Sars virus. Half a world away in a Singapore hospital, the victims fight for their lives. <Guardian April 3>

(61) [...] producing a logjam of critically ill patients who are now beginning to die, as it takes several weeks for SARS to claim its victims. <Time Asia May 12_3>

Since a war follows a certain pattern, the examples show that the discourse follows a similar course in a systematic manner. The bug has to *beaten*, before it *kills* its *victims* that *fight* for their lives. In the end, as in example (60), SARS will actually *claim victims*. This kind of lexical realisation of the discourse is not mirrored in the medical counterpart, and as a result, none of the words employed in the war metaphor can be found in the medical wordlists. The function of the conceptualisation in the popular texts is obvious. Picturing the disease like it is done in the medical texts means that the audience can develop a set of expectations about its nature and development. In addition, the news press has an easy-to-follow template at their disposition since the epidemic is pictured as an enemy that *attacks*, *kills* and needs to be *defeated*.

The group of collocates which includes *many*, *more* and *some* shows that *victims* is repeatedly used in a similar context as *people* or *patients* but with the difference that it implies to a fatal outcome.

(62) Many victims quickly develop severe pneumonia. <Guardian March 27>

(63) More than 80 victims in nearly two dozen nations had died of SARS by the weekend. <Time Asia April 14>

The function of *many* and *more* is similar to that identified for *people* as it is also used to give a vague, but somehow worrying interpretation of the number of people affected. Interestingly, the collocates *some* and *one* are used to refer to a positive process and seem to tone down the favourable development.

(64) Heymann said, some patients had finally started to recover, raising optimism that at least some victims can fight off the infection. <Washington Post March 27>

(65) There have been six cases of Sars in the UK, with just one of the victims still in hospital. <Guardian April 22>

Although the popular discourse on SARS seems to focus on the numbers of people affected and the cases confirmed, the key concerns were closely related to the scale of local threat. Example (66), for the collocate *new*, shows that other highly negative parallels are drawn.

(66) Forays into several hospitals in the provincial capital, Guangzhou, show that at ground zero new victims are still falling ill and dying. <Time Asia March 21>

Within the metaphorical system of "war and attack", the comparison of the area of the initial outbreak with *ground zero*, the terror attacks of 9/11 are fused with the SARS outbreak. This example is quite representative of the tone of some reporting throughout the crisis since, in 2003, the U.S.A. was also involved in the 'war on terror'. As a result, the two threats were blurred in the course of the coverage of the SARS crisis. The metaphorical reference to the terrorist attacks uses current anxieties and worries of the public to project them onto the epidemic. In doing so, a relation of equivalence is established that pictures the SARS virus as an animate entity that has a fatal power of mass destruction. Charteris-Black (2005: 162) analysed metaphors used in political discourse and confirms that terrorism is frequently employed to refer to the "myth of good and evil", where the public ("the good") has to face a threat that is presented in terms of metaphors ("the evil") that have a highly negative emotional appeal.

Lakoff and Johnson (1980) are convinced that metaphors not only make language more vivid and interesting but also structure perceptions and facilitate understanding for complex issues. Their view on the metaphorical structure of discourse is certainly relevant to the popular presentation of SARS, but I hesitate to declare the use of metaphors as a popularisation strategy since there is obviously no immediate link to aspects of the disease as such.

I will now proceed to the detailed analysis of the last candidate in Category II, i.e. *HCWS*, which is regarded as the most distinct item in the medical representation of SARS.

5.3.2.2 HCWS

HCW(s) is the top keyword in the medical subcorpus and is most obviously an alphabetism. The quantitative analysis in the previous chapter has shown that abbreviations and acronyms account for a considerable amount of the medical keywords. Specialised abbreviations enable efficient communication in a specific domain but are believed to pose a particular difficulty to comprehension outside the restricted field.

N	Concordance		
1	have been recommended for	*HCWs*	evaluating any person with
2	and probable SARS cases. All	*HCWs*	attending to patients with
3	suspected SARS (comprising 25	*HCWs*	and two inpatients; two
4	infection in 23 persons (11	*HCWs*	and 12 family members and
5	respiratory symptoms among	*HCWs*	, clusters of cases of
6	Appear to have infected > 10	*HCWs*	, family and social contacts
7	facility; of these, eight are	*HCWs*	and four are visitors. Of
8	body fluids. Of the 84 (42%)	*HCWs*	with probable SARS, 49 were
9	Two health care workers	(*HCWs*)	who were admitted to Tan
10	on educating and monitoring	*HCWs*	. Standard operating procedure

Figure 31: *Concordance sample for* HCWs *in the medical subcorpus*

Surprisingly, the concordance lines reveal that *HCWs* is actually an initialism for *health care workers* and does not stand for a special medical term like *CoV*. It is an acronym for a general multi-word expression that is just used in a specific way. This reflects the lexical routines of the scientific genre remarkably. *HCWs* is thus used in line with the prevailing concept of affected persons as they have been *infected* (lines 4, 6) and are *probable SARS cases* (lines 2, 8). Furthermore, they belong to a group of people that had been infected by patients that were spreading the virus to *visitors*, *family* and *social contacts* (lines 4, 6, 7). The numerical information that is provided alongside is also given for *HCWs*, but in a specific and unambiguous way by using precise figures like *84* or *42%* (line 8).

Table 46: *Collocates of* HCWS *in the medical subcorpus*

N	Word	Total	Total Left	Total Right
1	SARS	33	20	13
2	AMONG	24	23	1
3	CARE	16	12	4
4	FAMILY	15	2	13
5	HEALTH	15	14	1
6	TWO	15	8	7
7	CASES	14	10	4
8	INFECTION	13	11	2
9	PATIENTS	12	8	4
10	VISITORS	11	5	6

The collocations will serve as a starting point for the detailed discussion of these patterns and support the quantitative result in so far as the abbreviation *HCWs* is a key item in the medical subcorpus, but the component name parts in their full version are not. The following example furthermore reveals that in the rare cases where the abbreviation is preceded by the actual full expression, the context is rather general:

(67) Infections among health-care workers (HCWs) have been a common feature of severe acute respiratory syndrome (SARS) since its emergence. <MMWR May 16>

Examples (68) and (69), for the collocates *SARS* and *two*, give further evidence that whenever the abbreviation is used, it is also embedded in a specific context, or more exactly, co-text:

(68) On 2 separate occasions, ICU patients were transmitted SARS to HCWs in 2 of the 5 university medical-surgical ICUs in Toronto. <JAMA July17_2>

(69) Two additional HCWs (one each from hospital B and the outpatient laboratory) who had mild respiratory symptoms subsequently were furloughed from work. <MMWR April 25>

Apart from the ambiguous treatment of figures (*2* or *two*) the two extracts show an agreement in the co-text as specific expressions, such as *medical-surgery ICU* or *outpatient laboratory* are used which are particular to the medical texts. This is a consistent feature which is also observable in the following examples.

The collocates *patients*, *family* and *visitors* have to be interpreted jointly, since they are used in connection and point towards what actually happened to the HCWs that made them a key issue in the medical SARS discourse

(70) These patients appear to have infected > 10 HCWs, family and social contacts, or visitors to the health-care facilities where the patients were hospitalized. <WER May 9>

SARS appeared in hospitals where sick people had been admitted. The staff, unaware that a new epidemic had emerged, exposed themselves to the virus without using appropriate protection while they took care of the patients. These patients (sometimes also referred to as *index patients*) spread the virus as they were in close contact not only with medical staff but also with family members and other visitors. As a result, the number of SARS cases rapidly increased in March and April 2003. Example (71) suggests that the virus not only spread from patient to HCWs, but also, as the collocate *among* indicates, from health care workers to other medical staff:

(71) This report summarizes a cluster of SARS cases among HCWs in a hospital that occurred despite apparent compliance with recommended infection-control precautions. <MMWR May 16>

As this was an important matter, debates on this issue arose in the popular press and although they never seemed to go very far, their lexical realisation is significant. I have scanned a fragment of text in the popular subcorpus and, quite remarkably, the concordance lines reveal that two words that are regarded as key in the popular subcorpus, namely *doctors* and *nurses*, are used synonymously with the medical key HCWs. A very good prototypical example of this process is provided by the following extract from the Guardian:

(72) Many of those who fell ill initially were doctors and nurses who had treated patients with the disease. <Guardian April 2>

Thematically, the popular discussion was rather biased as it shifted from the mere statement that doctors and nurses died while caring for patients, to questioning whether this could have been prevented in the first place. Lexical profiling has shown a notable feature: a seemingly difficult medical abbreviation is actually a general expression that is just used specifically and there is nevertheless a popularisation process that replaces the abbreviation *HCWs* by the two hyponyms *doctors* and *nurses*, or, as the analysis of *health* in pilot study has shown, with its full denomination *health-care workers* (see section 3.4)

5.4 Category III: Words related

5.4.1 Related word forms: PATIENT(S) and DISEASE(S)

As already explained in Chapter 4.1, I opted for not lemmatising the corpus as there is sufficient theoretical and practical evidence that two word forms can have a significantly different distribution in a body of texts. The pilot study has already suggested that there is a substantial discrepancy between the use of the singular and the plural realisation of the lemma *patient*. The quantitative analysis for the entire corpus is in line with this as the plural form is also shared by both genres at high frequency, although it ranks significantly higher in the medical subcorpus and is thus seen as key. The singular form is even unique to the medical texts. The example of the lemmas for *disease* shows that this can also be the other way around as both corpora share the singular form with similar frequency, but the plural is seen as key for the popular subcorpus. Surprisingly, the plural *diseases* is particular to the popular representation of SARS although it is a semi-technical term that is taken from the medical domain.

PATIENT(S)

The plural form *patients* is shared by both collections of texts, yet it shows such a significant difference in relative frequency (0.94% in the medical subcorpus versus 0.22% in the popular subcorpus) that it is clearly regarded as key for the medical representation of SARS. Table (47) gives the five most recurrent lexical collocates for the plural form *patients* in both corpora.

Table 47: *Collocates of* PATIENTS *in both subcorpora*

N	Popular SC	Medical SC
1	SARS	INFECTIOUS
2	SOME	CARE
3	OTHER	HOSPITAL
4	HOSPITAL	RESPIRATORY
5	MORE	CLINICAL

The collocations indicate first of all that *patients* in the popular representation of SARS refers to an indefinite group of affected people since it associates regularly with vague quantifiers such as *some*, *more* and *other*. The study of the concordance samples in 3.4 has proved that no specific words or figures are given but indefinite quantifiers are used to make general statements, e.g. drugs have been proven ineffective in *some SARS-patients,* or, as example (73) illustrates, *other patients* did not respond to drugs:

(73) Other patients began showing symptoms of atypical pneumonia that were not responding to antibiotics. <Washington Post May 30>

If we compare this with the collocations for *patients* in the medical subcorpus, the difference becomes obvious as *patients* here associates frequently with modifying adjectives such as *clinical*, *infectious* or, as in example (74) also with *infected*:

(74) Clinical symptoms that were more common among infected patients who did not develop SARS than in patients with confirmed SARS were cough (72% of patients), sputum production (29%), […]. <BMJ June 21_3>

Example (74) also shows a salient feature, which is the precision of numerical information that co-occurs with *patients* in the medical subcorpus. Contrary to the popular texts, *patients* are specified since a definite number or, in this case, percentages are provided alongside. At the same time, patients are defined by their status or conditions, e.g. patients that are *infected*, patients that did not *develop SARS*, patients in whom a SARS infection is *confirmed*. The results show that the concept of *patients* is realised in both genres, but they are embedded in a different co-text and are used in different ways, i.e. with a general or a restricted meaning.

The singular form, *patient*, is regarded as key in the medical corpus and is used to refer to individual cases with a particular clinical presentation, which are sometimes part of a group of patients that share a certain feature or characteristic. This is illustrated in example (75):

(75) All of the patients were adults, ranging from 24 to 78 years of age. Six of the 10 were men. Eight of the 10 were of Asian descent. Three had a diagnosis of type 2 diabetes mellitus (Patient 1, Patient 2, and Patient 8); two had underlying pulmonary disease (asthma in Patient 3 and chronic cough of unclear cause in Patient 6). <NEJM May 15_5>

This extract is a summary of clinical features and case reports of patients that have SARS and makes use of anonymous, documented data. The particularity of the scientific context is reflected by the use of *Patient* (note the capital first letter) plus individual case number, for example *Patient 1*, *2* or *8*.

The findings support an interesting pattern since the newspress would not use the term *patient* for an individual, not because they do not distinguish between individuals – they normally do because personalisation is a highly important strategy in the mass media – but because there is a tendency to generalise, which is also reflected by the key status of *people* in the subcorpus. The wordlist analysis has yielded a number of proper names but they are predominantly health officials and are thus used in quotes.

DISEASE(S)

The discussion of the realisations of the lemmas for *patient* has shown that the more precise the context or reference is, the less likely it is that the usage of two related forms corresponds. This can also be a clue for the surprising fact that the plural *diseases* is only used in the popular representation of SARS, although the singular form is found with a close ranking (rank 3 in the popular subcorpus versus rank 12 in the medical subcorpus) in both subcorpora. Table (48) moreover shows a striking agreement of the most frequent collocates.

Table 48: *Collocates of DISEASE in both subcorpora*

N	Popular SC	Medical SC
1	SARS	INFECTIOUS
2	SPREAD	SARS
3	CONTROL	CONTROL
4	NEW	PATIENTS
5	INFECTIOUS	SPREAD

The subcorpora share not only the distribution of *disease* in terms of frequency but they also have four out of five collocates in common; only *new* and *patients* diverge.[27] Example (76), taken from the popular subcorpus, shows the collocates *new* and *spread* in context. Interestingly, it also shows that the relation between *SARS* and *disease* is of paradigmatic rather than of syntagmatic nature:

(76) No sooner had the early signs of severe acute respiratory syndrome (SARS) been detected, when it launched a well-publicized effort to combat the spread of the new disease. <Asia Times April 8>

SARS is introduced first and *disease* is used as a cohesive device subsequently. As shown in example (77), this pattern of paradigmatic substitution can also be observed in the medical texts:

(77) All hospitals must be prepared to care for patients with SARS. Thus, it is important to understand the transmission of this disease in hospitals. <CMAJ May 27>

This might account not only for the shared use of the collocate SARS but also for the overall frequency of the singular form in both corpora. However, using the hyponym as a means of maintaining cohesion and coherence in a text is a device that should not be judged in line with the collocational attraction of two words on the syntagmatic scale. This is in line with example (78) for the plural form *diseases* which in its high frequency is unique to the popular subcorpus:

(78) Because of the lingering controversy over etiology that has dogged other diseases, such as AIDS, it is worth going over the accumulated evidence that SARS-CoV indeed causes SARS. <Asia Times Sep 11>

[27] The reasons for the particularity of *new* for the popular and *patients* in the medical texts have been already been discussed in section 5.2.

Here, *diseases* is also used as an umbrella term for a variety of specific types. Once more, there is a reference to *other diseases* in a general way and only one is mentioned in particular for examplary purposes, i.e. *AIDS*. In choosing the denomination *AIDS* and not *HIV*, the causative agent is not specified as in the case of *coronavirus*, or *SARS-CoV*. The medical texts, on the other hand, prefer not only the expression *HIV* over *AIDS*, but also name the individual diseases that were referred to as *other* in example (79) taken from the scientific subcorpus:

(79) Compared with human immunodeficiency virus (HIV), tuberculosis, malaria, and even influenza, there have been relatively few cases of SARS and a limited number of deaths. <JAMA May 6>

The interesting feature is that a particular disease is not named for the function of exemplification, as in example (79), but to provide a consistent and complete record of related illnesses.

All in all, I am able to ascertain that there is a salient pattern which can be best described in terms of the dichotomy 'generalisation' in the popular discourse versus 'precision' in the scientific discourse. Remarkably, this accounts for both: a shared plural and a different singular form in the medical texts as well as a shared singular and a different plural form in the popular subcorpus.

5.4.2 Related words: FLU & INFLUENZA

As outlined in Chapter 2, the production of a popular scientific text involves a certain process of reformulation that modifies scientific information in such a way that it is no longer foregrounded and is thus comprehensible for a general audience. This can be done by employing different lexical and morphological devices to simplify a term, such as reducing its complexity by clipping. The simplified form presents an alternative to a specific item on the vertical scale. However, only the study of the syntagmatic level can reveal if the (apparently) synonymous relation holds completely true. The fact that the quantitative analysis has shown that the expression *flu* is unique to the popular subcorpus is in line with this since *flu* is the simplified, that is clipped, form of the medical term *influenza* which is likewise used in the medical texts. The question that arises and which should be answered by the contextual analysis is that of the status can be assigned to the word flu. Is it (1) a separate lexeme, (2) a genre-specific variant of a lexeme, or (3) two word-forms of one lexeme? Furthermore, it is most interesting to investigate if the two items are used synonymously. I will first take a closer look at the lexical profile of *flu* in the popular subcorpus (Figure 32).

N	Concordance		
1	of about 10, which is why	*flu*	-like epidemics are almost
2	Respiratory Syndrome (Sars), the	*flu*	-like disease which has
3	I was afraid because I had	*flu*	symptoms so my mother
4	international death toll from the	*flu*	-like Sars to at least
5	Countries with the mysterious	*flu*	-like disease. The EU
6	too lightly and warned the deadly	*flu*	-like disease could spread
7	spread of the potentially fatal	*flu*	-like virus. The Chinese
8	spread of the highly contagious	*flu*	-like disease. Worldwide
9	centuries earlier and the 1918	*flu*	that killed 30 million.
10	appreciate the dangers of bird	*flu*	and claimed to have it

Figure 32: *Concordance sample for* FLU *in the popular subcorpus*

The concordance sample already shows a remarkable pattern. When we look to the right of the node, *flu* is for the most part not used on its own but associates regularly with *like* and also frequently with *disease* (lines 2, 5, 6, 8). The resulting phrase relates SARS to the more familiar common cold and thus establishes a suitable comparison. The consistency to the left of the node is also striking in terms of semantic prosody. The adjectives that go with *flu* are highly negative and refer to *death* or *fatality* (4, 6, 7) or attribute a *danger* (lines 3, 6, 9, 10) to the *mysterious flu-like disease* (line 5). The only exceptions to the comparison with the common cold are those where other epidemics with fatal outcome are referred to, for example the outbreak of *Spanish flu* in 1918 (line 9) or *bird flu*. The collocates in Table (49) complement this picture:

Table 49: *Collocates of* FLU *in the popular subcorpus*

N	Word	Total	Total Left	Total Right
1	LIKE	77	10	67
2	SYMPTOMS	43	3	40
3	SPANISH	26	24	2
4	BIRD	21	19	2
5	DISEASE	18	6	12
6	VIRUS	18	4	14
7	PANDEMIC	16	6	10
8	PEOPLE	16	8	8
9	KILLED	14	0	14
10	SARS	14	5	9

Flu is used 249 times in total in the popular subcorpus and it occurs in almost one third of the cases in combination with *like*. The expression is either used to refer to a more familiar disease in general (*flu-like*), or as example (80) shows, to the *symptoms* of SARS that are similar to those of the common cold:

(80) The disease, which causes flu-like symptoms in patients, is now threatening the economy of Hong Kong and China, from where it first emerged. <Observer April 6)

There is a striking consistency as far as the semantic reference of the collocations is concerned, since *flu* is either used as a known reference point to which SARS can be related to in general, or it is used to refer to individual epidemic outbreaks:

(81) Cheng had just flown in from Hong Kong, where a father and son had recently died from the "bird flu," a deadly bug that experts long feared might spark a pandemic rivaling the Spanish flu of 1918-19, which killed as many as 50 million people. <Washington Post March 30>

Example (81) shows the relation between collocates. The most hazardous epidemics, or rather *pandemics*, as far as scope and fatality are concerned, are presented. Interestingly enough, *bird flu* has only claimed a moderate number of victims but has a history and outreach that is similar to *SARS*. In addition, *bird flu* is a disease that is still present and causes regular waves of outbreaks. *Spanish flu* has been contained for almost a century now, but a severe epidemic which killed a large number of people, especially in Europe. The common flu is referred to regularly as it is a known disease to which the public can relate. Furthermore, it has the potential to develop into a pandemic which draws an important parallel to SARS. The question arises whether this pattern will also hold true for the use of *influenza* in the medical subcorpus. The concordance lines in Figure (33) show that there are similarities as well as differences between the use of *SARS* and *influenza*.

Influenza is also used in connection with *bird flu* and other pandemics. In the medical texts the scientific designation *avian influenza* is given together with the specific virus type *H5N1* (lines 5, 7). When SARS is related to a possible *pandemic* scope, the uncertainty is stressed as there is already a level of *preparedness* although the pandemic has not yet taken place (lines 8, 10). Contrary to the popular subcorpus, there are concordance lines that refer to *influenza* not as a disease but as a *virus family* against which the strains of the SARS virus are tested and compared (lines 1, 2, 4, 9). The exception to this are presented in lines (3) and (6) where the disease is quoted in line with other severe infections, such as *tuberculosis*, *hemorrhagic fever* or *pneumonic plague*.

N	Concordance		
1	antibody testing against	*influenza*	A/B, parainfluenza 1/2/3
2	in vitro inhibition of RSV,	*influenza*	viruses and parainfluenza
3	with tuberculosis or	*influenza*	,certain viral hemorrhagic
4	antigen detection for	*influenza*	A and B, parainfluenza
5	associated with the avian	*influenza*	subtype H5N1, another
6	plague, adenovirus,	*influenza*	,mumps, parovirus B19,
7	influenza H5N1 and other	*influenza*	strains have been ruled out
8	that caused the incipient	*influenza*	pandemic in Hong Kong in
9	some of these reassortant	*influenza*	viruses are close to
10	information for baseline	*influenza*	pandemic preparedness.

Figure 33: *Concordance sample for* INFLUENZA *in the medical subcorpus*

So, there is no clear evidence if the two forms *influenza* and *flu* are used completely synonymously. However, the collocations for *influenza* in the medical subcorpus indicate that a number of collocates are related.

Table 50: *Collocates of* INFLUENZA *in the medical subcorpus*

N	Word	Total	Total Left	Total Right
1	VIRUS	22	5	17
2	H5N1	14	3	11
3	PANDEMIC	10	3	7
4	VIRUSES	10	0	10
5	AVIAN	9	9	0
6	RESPIRATORY	8	3	5
7	INFECTION	7	2	5
8	SARS	7	4	3
9	DISEASE	6	6	0
10	PARAINFLUENZA	6	0	6

If Table (49) and (50) are compared directly, there is direct agreement in four out of ten collocates (*virus*, *pandemic*, *disease* and *SARS*) and there is a further relation between the collocates *H5N1*, or *avian*, and the collocation *bird flu* in the popular subcorpus. Example (82) shows that the collocate *H5N1* is used in connection with other shared collocates, i.e. *pandemic* and *SARS*:

(82) A pandemic H5N1 virus (or any other subtype of virus) could be seeded globally in a matter of days, spreading at a faster rate than the SARS coronavirus. The similarity of clinical manifestations of recent H5N1 and SARS fatal pneumonia cases could pose problems in differential diagnosis. <Lancet May 10 2>

Contrary to the function of the collocates in the popular subcorpus, the explicit negative prosody is largely absent, although the possible hazard is outlined as the similarity of the diseases is stressed. However, the focus is on the spread of *bird flu* rather than on the spread of the *SARS* virus. The reference to the etiological agent has been a salient feature throughout this analysis and it is thus not surprising that example (83) shows, once more, a difference between the two corpora in terms of reference to a family of viruses. The collocates that are different from those of the popular subcorpus (e.g. *parainfluenza*) refer consistently to the comparison of agents:

(83) This family of viruses includes measles, mumps, human influenza and parainfluenza viruses, and respiratory syncytial virus in addition to the recently identified henipaviruses and metapneumovirus. <MMWR March 21>

The study of the two different realisations of the lexeme *influenza* has given evidence for some similarity as far as the associations with semantic sets are concerned, but has also drawn attention to the difference in lexical choice (e.g. *H5N1* versus *bird flu*). Furthermore, there are different functions that can be attested for the individual forms (e.g. the relation to a familiar concept vs. etiologic comparison) and there is, again, a clear negative prosody that associates with *flu* in the popular subcorpus (e.g. *deadly*). So, there is evidence for a popularisation strategy that places SARS in an explanatory context. Yet, there is also a difference in realisation that can be accounted for in terms of 'intention' on the medical side and 'attention' on the public side.

5.4.3 Related words: NASOPHARYNGEAL ASPIRATES & DROPLETS

So far, this section has investigated the contextualisation of words that are related in form. Now, the realisation of a shared important concept – the way in which the virus spreads – is a most interesting one since there is no obvious relation in form. Similar to the example of *HCWs*, there is an apparent process of transfer between the scientific and the popular discourse that takes the individual audience design and genre routines into account. *Nasopharyngeal* is key in the medical subcorpus and combines the concepts of *nasal* (referring to the nose) with *pharyngeal* (referring to the throat). It is thus morphologically complex and, with a length of 14 characters, likely to be perceived as a foregrounded item in general language use. The quantitative analysis in section 4.2 has suggested that a process of popularisation includes the reduction of (morphosyntactic) complexity. As the SARS virus spreads via the transmission of droplets, this concept should nevertheless be realised in both subcorpora .

First, Figure (34) presents the context for *nasopharyngeal* in the medical subcorpus.

N	Concordance		
1	coronavirus was positive on	*nasopharyngeal*	swabs from 4 patients.
2	and laboratory testing of	*nasopharyngeal*	aspirates and sera
3	Acute respiratory syndrome	*nasopharyngeal*	aspirates for virology
4	and the other from a	*nasopharyngeal*	aspirate of a woman
5	respiratory-tract fluids in	*nasopharyngeal*	aerosols have volumes
6	and the viral load in the	*nasopharyngeal*	aspirate to elucidate
7	We collected	*nasopharyngeal*	aspirates and clotted
8	onset of symptoms. The	*nasopharyngeal*	aspirates on admission
9	Also directly detected in	*nasopharyngeal*	aspirates from these
10	infectious agent, we took	*nasopharyngeal*	aspirates for virus

Figure 34: *Concordance sample for* NASOPHARYNGAL *in the medical subcorpus*

The examples show that *nasopharyngeal* predominantly collocates with *aspirates.* This cluster is used to refer to droplets that are spread via breathing and consist of *respiratory-tract fluids* (line 5) that might contain the *infectious agent* (line 10). If they do, they carry a *viral load* (line 6) and are subject to sampling and *laboratory testing* (line 2, 7, 10). The cluster *nasopharyngeal aspirates* is thus used in a quite specific setting and in connection with words that, once more, underline the importance of laboratory procedures. The collocates, provided in Table (51) confirm that, due to the orientation of the texts included in the medical subcorpus, this concept is certainly a key issue.

Table 51: *Collocates for* NASOPHARYNGEAL *in the medical subcorpus*

N	Word	Total	Total Left	Total Right
1	ASPIRATES	31	2	29
2	SAMPLES	21	5	16
3	ASPIRATE	18	1	17
4	PATIENTS	12	0	12
5	COLLECTED	9	5	4
6	PCR	8	5	3
7	RNA	8	7	1
8	VIRAL	8	6	2
9	RT	7	4	3
10	SWABS	7	1	6

The majority of texts in the medical subcorpus are research articles that are concerned with tracing the infectious agent of SARS as only the identification of the specific type of virus provides a solid basis for diagnosis and treatment. In the early stages of the disease, the type was still unknown and, as example (84) shows, samples of patients that have been fallen ill were obtained and analysed. In example (85) the coronavirus was already identified as the cause and tests now confirmed probable cases.

(84) Nasopharyngeal-aspirate samples obtained from all study patients were screened for common viruses. <NEJM May 15_4>

(85) Her nasopharyngeal aspirate was positive on RT-PCR for coronavirus and she seroconverted from titre less than one per 50 to one per 1600 to the coronavirus isolate. <Lancet May 24_3>

In example (86), the term *aerosols* and not the usual collocate *aspirates* was used. However, the cluster is preceded by another one that includes the word *droplets*. *Droplets* is a general word but is used, or rather contextualised, in a specific way.

(86) Droplets of respiratory-tract fluids in nasopharyngeal aerosols have volumes of 10-6-10-7 mL, so that the expected SARS virion population in a single droplet is between 0.1 and 10. <Lancet August 16_2>

Most interestingly the example also implies that, in order to substitute the term *aspirates*, a combination of both, *droplets* and *aerosols*, is required to refer to precise meaning. Since droplets is not supposed to be foregrounded in the popular text, the concordance lines (Figure 35) reveal that it is indeed used to discuss the spread of the SARS virus.

N	Concordance		
1	Sars has coughed and left	*droplets*	containing the virus
2	It is carried in small	*droplets*	, which means people
3	airborne rather than spread	*droplets*	from sneezing or coughing
4	and secretions such as	*droplets*	. You cough, a droplet
5	is believed to be spread by	*droplets*	from sneezing and coughing
6	to person through airborne	*droplets*	- a sneeze or a cough -
7	is still believed to be in	*droplets*	from the nose and mouth
8	than the relatively large	*droplets*	expelled in sneezes and
9	contact with pathogen-laden	*droplets*	, the other is fear, which
10	primarily through tiny	*droplets*	expelled by infected

Figure 35: *Concordance sample for* DROPLETS *in the popular subcorpus*

The sample also shows that there is not only a direct relation to the concept of *aspirates*, but even more so to *nasospharyngeal.* The term is split into its two lexical morphemes *nasal* and *pharyngeal* which are then substituted by the respective familiar terms of vernacular origin; *nose* and *mouth* (line 7) or even more exactly *sneeze(ing)* or *cough(ing)* (lines 3,4,5,6,8). In line 6, there is also a reference to *airborne* which can be related to *aerosols* in the medical corpus. All in all, the conceptualisation is similar to that of its scientific counterpart although in the popular representation, as the collocates in Table (61) show, the concept is embedded in a verbal environment that aims at explaining how the virus spreads and not how it can be identified in the first place.

Table 61: *Collocates of* DROPLETS *in the popular subcorpus*

N	Word	Total	Total Left	Total Right
1	SPREAD	13	13	0
2	AIRBORNE	7	3	4
3	COUGH	7	1	6
4	INFECTED	6	4	2
5	PERSON	6	5	1
6	SARS	5	3	2
7	TINY	5	5	0
8	VIRUS	5	3	2
9	SNEEZE	4	0	4
10	SNEEZING	4	0	4

There is a distinct group of collocates for *droplets* that characterises and paraphrases the concepts of *nasaopharyngeal aspirates*. They are *airborne* and *tiny*, they *spread* through a *cough* or a *sneeze* and, if they contain the *virus*, they *infect* a *person* with *SARS*. The following example (87), taken from a lead article in the Guardian at an early stage of the disease, shows how this is realised in the popular texts.

(87) Someone with Sars has coughed and left droplets containing the virus, and then the next person gets these droplets on their fingers, touches their eyes or mouth and becomes infected. <Guardian April 6>

The example also shows that the context is predominantly an explanatory one and the reference to the laboratory procedures that prevails in the medical texts is absent. This is certainly due to the different communicative objectives of the target audiences. The general public wants to understand how the virus spreads and what precautions they might take to prevent an infection. As this is the overall aim, comprehension and understanding is fostered by reducing complexity and using familiar concepts that are realised though the use of general or institutionalised terms such as *virus* and *infection*. Example (88) confirms that *droplets* is consistently embedded in such an explanatory context.

(88) As far as health officials know, SARS is spread by tiny droplets that get airborne when someone sneezes or coughs, or by contact with other bodily fluids, such as blood. <Washington Post April 24>

To frame the topic, officials, or more exactly health officials, are referred to as they have obviously passed the information on. So, there is substantial evidence for a process of popularisation since a specific concept is transferred from the medical circles to the general public. For doing so, the morphological complexity of the term was reduced as general lexical items were chosen to refer to the idea of respiratory droplets containing the virus and the concept was recontextualised and embedded in a general, explanatory context. The contextualisation is similar to the realisation of *coronavirus*, although this time, the medical expression is not introduced but is completely substituted. There is a relation to the lexical process that was involved in the realisation of the abbreviation HCWs that was substituted by doctors and nurses in the popular texts. Yet, again, the lexical choice was different but the contextualisation remained similar. All in all, there is a parallel that can be drawn as the findings have proven that if a process of popularisation takes place, negative semantic prosody is largely absent.

5.5 Summary

In this chapter, the focus has been on the way in which words behave in their particular settings and on how certain prominent words are handled in the different genres. Although the quantitative analysis has suggested an interrelation between the two subcorpora, the qualitative analysis has clearly shown that the words in use are highly context-dependent since they derive much of their meaning from their associates. However, I am now able to ascertain that no firm divide can be established between the scientific and the non-scientific world, although a description in terms of a scale with two extreme ends can be made. The study of concordance fragments and co-occuring words has shown that there are recurrent patterns that are specific to the particular genres which can be interpreted according to:

- *Style*: general vs. scientific (e.g. *flu* and *influenza*),
- *Intention*: explain vs. inform (e.g. *coronavirus* and *SARS Cov*),
- *Function*: generalise vs. specify (e.g. *people* and *patient*),

- *Appeal*: newsworthiness vs. practical implementation (e.g. *spread* and *containment*).

One of the major conclusions that can be drawn, also from a methodological point of view, is that words that are shared by both genres frequency-wise (Category I) can be used differently (e.g. *virus*), whereas seemingly different words (Category II) can show a systematic agreement in their lexical profiles (e.g. *HCWs*). The lexical profile is of course an abstraction but has proved useful for the characterisation of a candidate word as far as collocation and semantic prosody are concerned. All in all, there are related genre-specific prosodies, but they can be interpreted more in terms of association with semantic sets than with negative or positive semantic prosody. Negative prosody is employed only in the popular texts and adds a certain bias or evaluative sense to the words. As a result of this analysis, semantic prosody can show at least two significant things: the ways in which general and specific linguistic items are recontextualised and the way words become more collocationally fixed in the specialist environment. There is a preferred way of expressing a delaminated set of semantic reference in both genres but I would argue that it should be regarded as a sometimes unconscious system of preferred expressions differentiated by the rhetorical aims of the discourse communities (e.g. precision vs. complexity reduction). The discussion of SARS should therefore not be seen as a body of facts transmitted via a linear popularisation process but rather as a special linguistic construct which is mediated by recontextualisation.

6 Towards a model of lexical processes

In the two previous chapters, the lexical interrelation of the two genres was analysed from a quantitative perspective, focusing on the distribution of single words, and from a qualitative perspective which extended the scope of the analyses towards the verbal co-text of these words. The qualitative results confirm that, in the SARS discourse, lexical meaning is not only constructed by the use of particular terminology but is first and foremost shaped by its individual co-textual and contextual variables. The theoretical discussion in Chapter 2 has suggested that the realisation of a scientific topic, such as SARS, entails particular elements like the verticality of discourse and a particular audience design which depends on the readers' expectations and ability to employ the resources available in the register. The analyses have also proved that the lexical realisation is, to a great extent, influenced by the individual genre routines. The present chapter is intended to place the results of the corpus-based analysis within the theoretical framework of scientific discourse popularisation. To this end, systematic lexical patterns will be described (section 6.1), a model of lexical processes will be proposed (section 6.2) and the corpus data will be joined with theoretical considerations to add to a linguistic theory of popularisation (section 6.3).

6.1 A description of parallel patterns and individual characteristics

The different forms of the SARS discourse can be traced down to a common reference point, i.e. the disease and the events associated with it, but they still circulate in quite distinct, separate settings and employ different concepts to describe the issue. The quantitative categorisation of the most common words used in both subcorpora has given a good indication for lexical overlap. However, from a sheer quantitative perspective, there is as much evidence for common ground as there is for individual traits. The qualitative corpus analysis has helped to refine initial results in tracing patterns of contextual use which give a first indication for certain processes of lexical realisation within the texts. In doing so, particular patterns and genre routines (i.e. the use of quantifiers instead of figures, the use of epistemic modals) can become obvious. So, as Mukherjee (2002) asserts, corpus-linguistic methods are an effective tool to trace and describe such patterns and routines:

> Es zeigt sich einmal mehr, dass [...] Bedeutung immer auch durch sprachliche Routine geschaffen wird: Mit korpuslinguistischen

> Methoden kann man dieser Routine auf die Spur kommen und zu wertvollen interpretatorischen Schlussfolgerungen gelangen. (Mukherjee 2002: 137)

Before a model of lexical processes in scientific discourse popularisation can be sketched out, it is necessary to trace routines in order to describe patterns that give evidence for a process. In doing so, this section sets out to answer the research questions regarding lexical choice and the relation between form and function (as outlined in Chapter 2.4). A central point is the issue as to whether there is a related lexical realisation of a scientific topic in the two genres and how potentially shared items function in the individual settings. The quantitative analysis has shown that almost half of the very frequent content words are actually shared by both corpora. So, there is evidence of common ground between the scientific texts and the general news articles. In order to pinpoint the degree of interception between the two distinct levels of discourse, characteristic lexical patterns can be categorised: (1) semantically on a paradigmatic level (i.e. semantic fields), (2) semantically and lexically on a syntagmatic level (i.e. prosody in terms of co-selection of semantic sets and connotations), (3) from a lexical and stylistic perspective (i.e. the use of general words, semi-technical expressions, specific terms or metaphors), as well as (4) from a lexical and functional perspective (i.e. reduction of complexity vs. precision).

Semantically, the fields that have been established in Chapter 4 are still valid, although the contextual study of the individual items has given evidence for genre-specific variation on the paradigmatic as well as on the syntagmatic level within the categories. The study of the field '*persons & institutions*' is of particular interest as it shows that not only foregrounded items, like complex scientific terms such as *nasopharyngeal aspirates*, are subject to re-contextualisation but that general concepts can also be differently contextualised or primed. Within this semantic field there is reference to persons that are directly affected by SARS as they have fallen ill with a possible fatal outcome, i.e. *patient(s)*, *cases(s)*, *victims*, then there is reference to those that carry the virus or are at risk of getting infected (*people*, *HCWs*) and to those who are professionally involved in the management of the disease *(scientists*, *spokespersons*, the *WHO)*. The contextualisation of the concepts and the preference for certain choices that can be made when referring to persons that are (potentially) infected is depicted in Table (62). It also reports on the most interesting pattern of the co-selection of quantitative information that adds to the construction of a conceptual unit and could be considered as an indication of lexical priming.

Table 63: *Realisation of the concept* affected persons

	POPULAR		MEDICAL	
	use in texts	/ co-selection	use in texts	/ co-selection
case	no	_	yes	(individual case number, e.g. *case 3*)
cases	yes	(adjectives, e.g. *suspected cases*)	yes	(adjectives, individual case numbers, e.g. *probable cases*)
patient	no	_	yes	(individual numbers, e.g. *1 patient, index patient 3*)
patients	yes	(quantifiers *some patients*)	yes	(concrete, small figures e.g. *8 patients*)
people	yes	(global figures or quantifiers, e.g. *2,890 people*)	no	_
victims	yes	(quantifiers, e.g. *many victims*)	no	_

It is a striking fact that the singular forms, *patient* and *case*, are unique to the medical texts whereas the popular texts make use of plural forms, or the collective noun *people*, only. On a scale with two extreme ends, the co-selection of numerical information would commence with individual numbers that collocate with *case* and increases towards the use of global figures or indefinite quantifiers that collocate with *people* and *victims*. So, *cases* are primed to be used with adjectives that indicate likelihood, *patients* to be used with rising numbers and *people* triggers vague quantifiers or global figures. In the medical texts, the respective lexical choice is also tied to numerical information but it is given in a more precise form by exact figures and/or relative percentages. This semantic field is thus of particular interest since it not only provides insights into the divergent lexical choice but furthermore gives evidence of the process of priming. In Hoey's (2005) terms, the systematic co-selection of figures or quantifiers supports the fact that the different realisations of the concept is subject to lexical priming as the different choices trigger particular collocates.

The theory of priming, however, cannot account for obviously functionally motivated differences in the use of other shared terms. The second prevalent concept of individuals and organisations involved in the management of SARS, for example the *World Health Organisation*, does not so much differ in its co-textual selection but rather in its contextual function. Apparently, there is a preference in the popular texts to not only refer to the WHO as the coordination institutions, in the same way as in the medical texts, but to use official organisations to place the author's own evaluation and predictions on reliable ground. Example (89) shows how this is employed and contextualised in the popular texts:

(89) Officials warn of virus's ongoing threat. SARS could recur anywhere at any time. <Washington Post May 1>

The embedding in negative prosodic contexts is a clearly genre-specific pattern as journalists usually draw on prestigious, that is, officially acknowledged, sources to use them for the attention-seeking construction of a story. Bell (1996: 221) claims that in the course of this process "boundaries between who said what, and what is the journalist's own input get regularly blurred". The corpus study has shown that direct and indirect reference to the WHO, its affiliated scientists and spokespersons is a communicative means that is extensively employed to back up evaluations and views on the epidemic in the popular texts. This highlights the functional importance of the category *comments* in van Dijk's (1988) model of news structure, as illustrated in Figure (36).

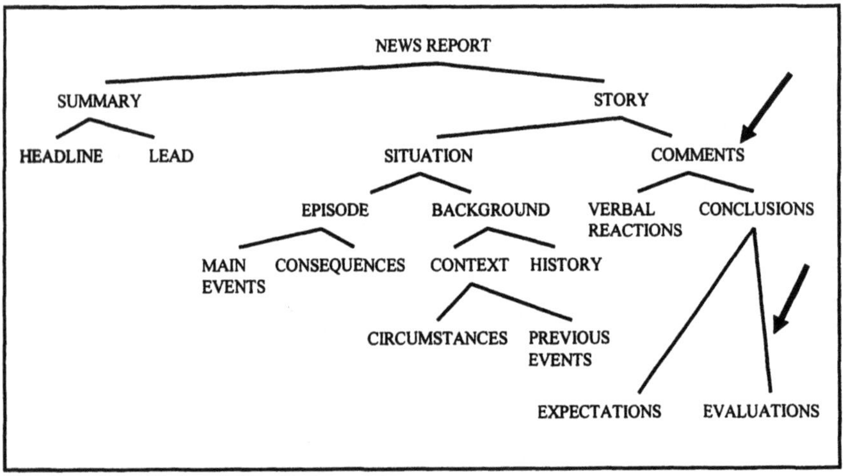

Figure 36: *Superstructure of news reports: comments and evaluations*

Comments include verbal reactions of officials and there is a distinct group of proper names (*Stohr*, *Gerberding*, *Heymann*) that are particular to the popular texts and the names of the individual experts quoted are shared. Calsamiglia and Lopez Ferrero (2003) have investigated the use of citation and reported speech in discourse popularisation. They claim that journalists use quoted speech as a means "of abdicating the responsibility of journalists or a means of orientating their position on the topic of reference. In doing so they dramatize the conflict". (Calsamiglia and Lopez Ferrero 2003: 169). By adding relevant verbal reactions, the journalists try to obtain higher credibility for evaluations and predictions. This is, however, only objective to some degree as they offer ready-made interpretations which reflect their own opinion and express (though not on purpose) a certain stance or negative bias.

Coming back to the categorisation of shared semantic fields, the two realisations of the SARS discourse refer repeatedly to *place & tim*', although this field is obviously more dominant in the newspress. It includes mostly proper names of the areas where the virus has spread from (*Asia*, *China*, *Guangdong*), where it has spread to (*Hong Kong*, *Canada*, the *world*) and the time when it spread (*March*, *April*, *May* 2003). In both realisations, this is embedded within a general context that is related to the outside event but not to the medical framework of SARS. So, whereas the first two categories are of rather a general nature and do not employ specific medical terminology, the third common field, the *cause*, *agent & symptoms of SARS* centres around scientific concepts and makes use of institutionalised (semi-technical) terms, such as *virus*, and of specific terminology, such as *respiratory*. The keyword analysis has shown that, in the medical texts, the individual words that are supposed to give an indication of the 'aboutness' of a corpus, can be largely grouped into this category. The scientific concepts that are fundamental to research articles but do not have a corresponding realisation in the general newspress refer to diagnosis and treatment of the disease as well as to laboratory features and genetic sequencing of the virus. Differences between the concepts that are shared (e.g. identification of the pathogen) involve abstraction and precision, as there is a preference in the popular texts to use the institutionalised, superordinate variant (*virus*) if the corresponding medical concept is at a lower taxonomic level (*metapneumo-virus*). Consequently, a popularisation process takes predominantly place within this category. If the terminology that is not institutionalised is indeed transferred, a process of assimilation takes place; *coronavirus*, for example, is embedded in an explanatory context.

The last shared semantic field *spread, countermeasures & outcome* is of particular interest since here general words are used in a specific way. The corpus analysis has shown that there is a distinct group of words, such as *contact, control* or *isolation*, that undergo a process of semantic restriction or shift in order to fit into the respective contextual frame. *Isolation*, for example, is used synonymously with *quarantine* (of infected people) in both genres, but it is used to refer to the isolation of genetic sequences in the medical text only. The words are general in terms of form but specific in terms of contextualisation as they may have different functions in different discourse settings.

This semantic categorisation provides a scaffolding for the patterning, but to make a decisive statement about lexical processes, the individual lexical and stylistic choice within the fields has to be depicted systematically. The model proposed in Figure (37) illustrates variation within the seemingly homogenous group of shared words and takes account of the different levels involved in the realisation of common lexis.

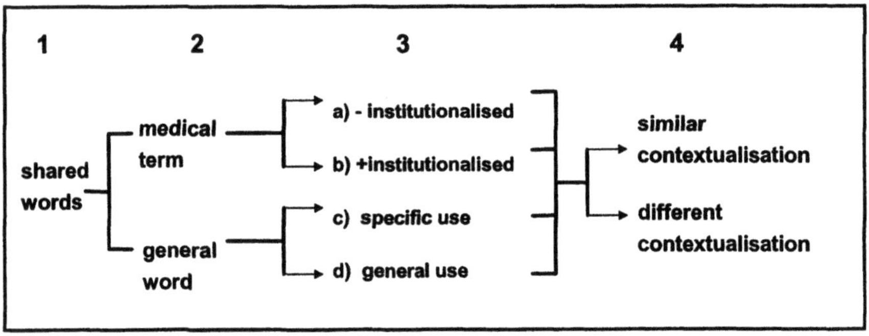

Figure 37: *Use and contextualisation of shared words*

The model draws on shared words (1) or, more exactly, on shared word-forms as identified in Chapter 4, and (2) makes a distinction between words that are common in everyday-language use and terminology used in the medical domain. The third layer (3) takes account of the divergent status of those items. On a vertical scale, which ranges from LSP to general language, a) is the highest level of specialisation and includes terminology that is particular to scientific language use; b) includes terms from the medical domain that are institutionalised, and thus to some extent also used in general language, while c) and d) refer to words that are not subject-specific but of a general nature and are either used in a restricted way (c), or are not specific at all but in accordance with the common usage norm (d). Finally, in (4) the difference in contextualisation, which can be functionally motivated or is an outcome of lexical priming, is considered.

The objective of this model is to show that identity in form (1 and 2) does not automatically entail identity in use (3 and 4). A word or lexical item, such as the acronym SARS, may be used similarly frequent in both corpora.

At the next level (2) it is characterised as a medical term which is, at first, particular to LSP (3a) and was later in the course of the coverage institutionalised in general language use (3b). This is marked by the use of the acronym without its component parts (see section 5.2). As far as contextualisation (4) is concerned, SARS collocates in both corpora with similar semantic sets and clusters other than the name components (e.g. *SARS virus*, *SARS cases*, *transmission* or *spread of SARS*). In the popular corpus, however, it also co-occurs with negative prosodic sets such as *warning* and *fear*. This might be an indication for an ongoing process of institutionalisation, since terms in category 3a), that is, those that are not yet institutionalised, may just show a different contextualisation in terms of semantic sets, but not in terms of negative prosody. *Respiratory*, for example, is obviously used as part of *severe acute respiratory syndrome*, but is also used independently in the two collections of texts as it is refers to the type of *infection* or *illness* in the popular corpus, while it denotes the specific type of *virus* or *specimen* in the medical corpus. The example of *coronavirus*, which is regarded as key for the medical realisation of SARS but is also used in the newspress, shows how a term that is alien to the general public is subject to a process of transfer. It is embedded in an explanatory context that makes use of the same semantic sets as its scientific counterpart, i.e. by drawing parallels to other members of the coronavirus family. However, there is once more a difference in the individual lexical choice since the two realisations stay at a different level of a tentative taxonomy (e.g. *common cold* vs. *human coronavirus OC43*).

This example further supports the observation that *coronavirus*, in contrast to the institutionalised term *virus*, shows no recurrent negative prosody but rather an association with semantic sets that facilitate comprehension. *Virus*, and also *patients*, thus belong to category 3b), which embraces terms that originate from the medical domain but are institutionalised and widely used in general language. The two words are differently contextualised although *virus* co-occurs with highly negative prosodic words while *patients* is differently primed in terms of different realisation within a particular semantic set. In the medical texts, the context of *virus* is highly specific and restricted as it is consistently used in multiword expression to provide precise reference (e.g. *bovine*, *hepatitis*, *syncitial virus*). In the popular corpus, *virus* collocates with words that have a distinctly negative connotation such as *deadly*, *killer* and *worries*. Surprisingly, this negative prosody is largely absent in case of shared general words (3c&d). General words in 3c) are everyday terms that are particularly used in the SARS discourse and are embedded in a specialised or restricted context. The findings from both the quantitative and the qualitative analysis highlight words that are shared between the genres, such as *health* or *contact*, but deviate from the general usage norm. *Health* is used predominately in multiword clusters to refer to organisations and professionals (e.g. *health-care professionals*) and not to denote a physical state. *Contact* is not only used to refer to interpersonal or *face-to-face* contact, but is, in the medical texts, also transferred to a laboratory setting as it collocates regularly with *specimens*. The example of *world* shows that the two levels of the SARS discourse share its specific (3c) but also its general (3d) use. It is either used specifically as part of the name *World Health Organisation* or it refers to the global span of the disease. This is in line with the observation that in 3d), general words that belong to the '*place & time*' category are neither particularly primed nor do they fulfil a genre-specific function.

As far as individual traits are concerned, the different contextualisation of shared words highlight systematic differences in lexical choice (general vs. precise terms), semantic prosody (co-selection of negative items in the popular texts) and lexical priming (number vs. quantifiers). The quantitative analysis has shown that there is as much evidence for individual words as there is for common ground. The use of individual forms is either due to the realisation of a semantic field that has no correspondence in the counterpart (genetic sequencing, diagnosis and treatment) or is due to the choice of rhetorical aims (metaphors, abbreviations). The corpus analysis has shown that some different forms can indeed be related, which gives evidence of a popularisation process as the lexical choice is a functional device to reduce complexity in the texts written for the general public. The examples of the lexical realisation of *nasopharyngeal aspirates*, as well as the substitution of the abbreviations *HCWs*, has given evidence for a process that Gotti (2005) compares with "intralinguistic translation":

> The popularisation process is kind of redrafting that does not alter the disciplinary content – object of the transaction – as much as its language, which needs to be remodelled to suit a new target audience. In the process, information is transferred linguistically in way similar to periphrasis or intralinguistic translation. (Gotti 2005: 205)

The study of the different distributions of mean word lengths in section 4.2 has already suggested that the morphological complexity is reduced in the popular texts and the qualitative analysis has provided evidence for this. The language barrier between the scientific and the public audience is lowered as the complexity of scientific information is reduced in four ways:

- reducing the density of scientific concepts by selection of information that is immediately relevant to the target audience;
- morphosyntactically complex constructions are paraphrased by the use of more general words (*nasopharyngeal aspirates* = *droplets from coughing and sneezing*);

- foregrounded items are simplified, for example by clipping (*influenza* = *flu*);
- genre-typical acronyms and abbreviations are given in full or are paraphrased (*HCWs* = *health-care workers*, *doctors and nurses*).

The case of *SARS* has shown, however, that as soon as an acronym is institutionalised in the public discussion of the topic, the complete denomination is no longer used and its component parts might even be opaque for the users of the acronym.

Individual traits that are not transferred are, on the one hand, the overall density and precision of the medical texts indicating a scientific functional dialect and, on the other hand, the stylistic and rhetorical devices of negative prosody and metaphors associated with that in the newspress. The metaphors and negative polarity items, such as *victims* and *killer virus*, used to frame the epidemic in the popular press, have a crucial role in demonstrating the potentially fatal outcome of the epidemic and thus shape the impact of disease on the general public. The conceptual metaphors that are employed (*war* and *terror*) express a certain attitude and stance to highlight the narrative 'story' character and thus newsworthiness of the topic. This should clearly be considered a genre routine and not as evidence of a transfer process. All in all, the description of shared patterns and individual traits has shown that it is important to distinguish between the individual preference to express similar concepts or to realise individual concepts and routines in the first place. Lexical choice and contextualisation of common ground is determined by the appropriateness in the given situation and genre.

6.2 A dynamic model: framework and constituents

The description of lexis in the dual representation of the SARS discourse has given evidence for transfer of information as well as for independent lexical realisation of the topic. I would therefore subscribe to the view that the discourse takes place on a continuum and that lexical processes within this continuum have to be described by relating the internal linguistic properties of texts to their external variables. This is in line with Baumann's (1992) argumentation that without the contextual characteristics no adequate statement about popularisation can be made: "Der Grad der Fachlichkeit ist vielmehr das Ergebnis eines komplexen Zusammenwirkens von außersprachlichen, innersprachlichen und funktionalen Faktoren" (Baumann 1992: 153). I have already outlined the basic factors for a theoretical framework of the SARS discourse (see section 2.3), but in the light of the corpus analysis the preliminary representation of verticality should be refined. This is a prerequisite for a comprehensive model that maps lexical processes within a particular theoretical and situational framework. The interpretation of results has to consider norms of a community which shares certain values and preferred lexical structures which are obviously expressed in particular rhetorical choices. The extended model of the verticality of the SARS discourse, depicted in Figure (38), includes five levels.

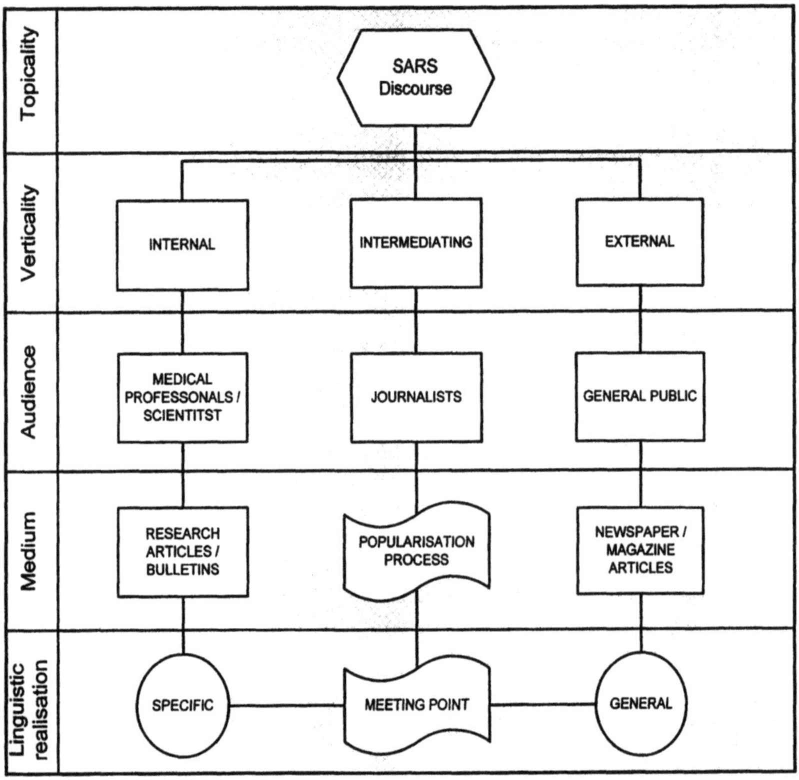

Figure 38: *Five levels of popular discourse*

To put popularisation into a framework the model considers (1) the overall topic, (2) the verticality of discourse, (3) the audience that constitutes the particular discourse community, (4) the medium that is used, and, (5) what is actually visible, that is, the linguistic realisation with the language in use. Each of the different factors within the levels should be regarded as a crucial variable since the replacement of one variable with another has an effect on the entire realisation:

(1) Topicality: topic-bound factors influence the lexical choice since SARS, for example, requires the reference to medical aspects as well as to features that are unique to an emerging disease with a pandemic scope. This includes the importance of disease control by taking precautions and by identifying the pathogen, as well as the impact of the disease on the world in times of global mobility. As a consequence, the linguistic realisation makes use of general vocabulary (to refer to *place & time*), general words that are used in a specific context (to discuss the containment of the disease), as well as of specific terminology (to describe symptoms and the pathogen).

(2) Verticality: factors concerning the vertical division of the discourse have to distinguish between internal communication that takes place within the limits of the special field, intermediating communication which passes on information from inside the domain to the outside, and external communication where a topic is discussed completely outside the realms of its special field. Popularisation takes place at an intermediate (and later external) level. One has to be aware, however, of the plurality within the intermediate stage as it is essential to make a distinction between the different modes of popularisation. Knowledge transfer for didactic purposes cannot be analysed in terms similar to the news coverage in the mass media since it is certainly closer to the internal level than it is to the external.[1]

(3) Audience: reader-bound factors are tied to verticality since audience design should obviously differ from level to level. Yet, this straightforward concept is not as easily applicable as it may seem at first sight, because lexical profiling has shown that audience design should not only consider the subject-specific knowledge but also the text-typological knowledge of a target group. If information about a scientific topic is passed on, the journalist, as a mediator, has to negotiate between the demands and preferences of experts that have shaped the information in the first place and the language use that is familiar to the general public. The corpus analysis has shown that this might result in a duality of expressions. Precision and economy, for example, is required and expected by the readers of the medical texts (e.g. choice of items from a low taxonomic level; use of abbreviations), whereas generalisation and elaboration of facts facilitates comprehension in the popular texts (e.g. choice of items from a high taxonomic level; words instead of abbreviations).

---

[1] See also Wichter and Antos (2001).

(4) Medium: medium-bound factors include first of all the mode (oral or written) and, subsequently, the particular text type. For the SARS discourse, the media for internal communication are research articles and bulletins that the WHO has passed on to the respective medical journals. As the intermediating level has no manifestation, the actual outcome of the popularisation process manifests itself at the external level in form of newspaper and magazine articles[2]. There is, of course, also verticality within the newspress as one has to make a distinction between upmarket papers that aim at informing the readers and tabloids that might more aim at entertaining their audience. So, differences in the lexical realisation of the topic are also due to the routines of the individual media. While research articles are a platform to present and discuss findings by employing numerical data and individual case numbers, the popular texts show almost uniform distribution of quantifiers and global numbers. The newspress, on the other hand, uses conceptual metaphors and a variety of items that have a distinct negative prosody which is completely absent from the medical texts.

(5) Linguistic realisation: this is the level where the register-bound factors (verticality and audience) and the genre-bound factors (verticality and medium) are visible. The last stage is the only one that can be observed and analysed to get an insight into the lexical choice that has been motivated by the previous stages. In scientific discourse popularisation, it is the meeting point for specific and general language. As this level is the most complex one, this is where a model of lexical processes should be added.

So far, systematic patterns in terms of lexical similarities and deviances as well as a conception of the particular situational framework have been outlined. What is missing is the link between the two approaches and the model proposed in Figure (31). Therefore, the comprehensive model outlined in Figure (39) joins the new concept of verticality with lexical patterns in scientific discourse popularisation. It takes the following essential constituents and variables into account: topic or subject-matter, level of verticality, discourse form, lexical choice that leads to the co-selection of contextual items, resulting contextualisation which might be influenced by genre-specific functions or priming and links form and function to the particular genre routines.

The first part of the model is a condensed form of the previous framework that pictures the crucial elements of scientific discourse popularisation. As we have seen already, the topic can be realised at different levels on a vertical scale.

---

[2] This obviously depends on the level of popularisation. Knowledge transfer would have a manifestation at the intermediate level, for example, in the form of educational text books.

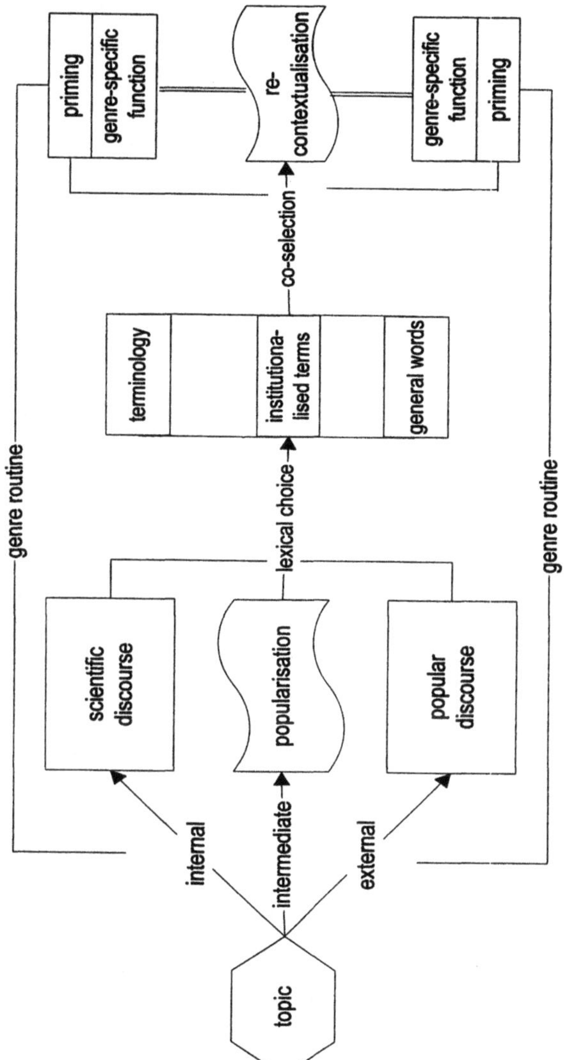

Figure 39: *Lexical processes in scientific discourse popularisation*

From a lexical perspective, verticality cannot be applied to the external, communicative setting only but also to internal features like word-choice and vocabulary ("Vertikalität des Wortschatzes", cf. Busch 1994, Steger 1988, Wichter 1994).This concept describes a gradience with two opposing points: terminology and general language. The analysis has shown, however, that the actual lexical choice takes place in-between the one and the other extreme end of the scale. In addition, there is also a point in the middle of the continuum where the two discourse forms meet and where boundaries are blurred. When terms that belong to a specific domain are taken over by general language use on a regular basis, they get institutionalised and are used independently in a general context. In reality, there is a certain focal point on the scale where, from a quantitative point of view, the most items that are used can be pinpointed. The scientific discourse is closer to the 'terminology end', whereas the popular discourse tends towards the 'general end'. However, the preference for one of the ends does not mean that the two discourses do not exploit the full range and make use, to some extent, of all three possibilities of lexical choice. The co-selection of items is certainly governed by the particular lexical choice and the resulting re-contextualisation is immediately related to the popularisation process and draws on the patterns described in 6.1. I prefer the term *re-contextualisation* over *de-contextualisation*, since the examples that give evidence of a popularisation process have shown that the term is not used outside of the context but just in an adapted kind of it, i.e. a more general or explanatory one as is the case with *coronavirus*. Those terms that are used in a different context can be traced in the middle of tentative scale where institutionalised terms such as *virus* are located. The realisation of *virus* is a most interesting one since the systematic pattern of the designation for the causative agent of the disease reflects the vertical and horizontal relationships among the different choices. This is influenced by the genre routine which is immediately related to re-contextualisation and priming. Figure (34) depicts a taxonomy of the hyperonym *virus* and its hyponym *SARS-CoV*. In this taxonomy of the SARS virus, the popular and the medical texts make common use of the concepts that are located at the two upper levels (*virus*, *coronavirus*) and the lowest level (*SARS-CoV*) but only the medical texts exploit the range to the full and refer to the levels inbetween. In the light of popularisation, two of the basic strategies are the reduction of foregrounded items, such as complex terminology, and the elimination of information that is not relevant to or cannot be readily interpreted by the audience. As a consequence, only the basic term *virus* and (SARS) *coronavirus* are used frequently in the popular text. However, the levels should not be mixed up because, as Lipka (1992) argues, the substitution of an archilexeme with its subordinate term will falsify the truth value of an utterance (i.e. *virus* can only be substituted by *coronavirus* in specific contexts).

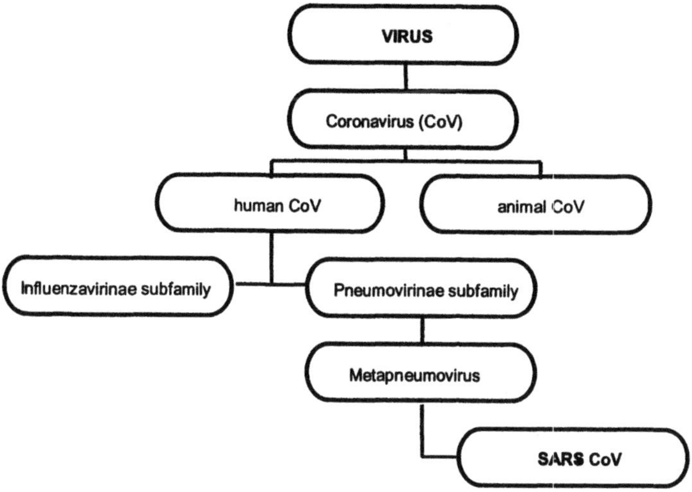

Figure 34: *Taxonomy for VIRUS*

All in all, I would argue that the lexical choice of an item from a certain level of the taxonomy is more related to verticality than to popularisation. Most interestingly, terms that are institutionalised in general language use undergo a particular process of priming to specify or split their general reference in everyday use from the specific meaning they take over in LSP. Consequently, *virus* is consistently modified by nouns and adjectives in order to ensure a precise reference. This is one of the basic mechanisms of priming in scientific discourse and the pre-modifiers take on, or are primed for, a designative or categorising function.

In contrast to this, the realisation of a certain genre-specific function in the particular genre includes, for example, the use of complex, dense terms in the scientific realisation and the use of metaphors or negative prosodic items in the popular one. One of the main differences between the discourse forms is that of semantic prosody. Bell (1996: 216) asserts that this is a particular trait of the newspress since "news values drive the way journalists write their stories" and the newsworthiness is high if an immediate (negative) effect on the reader is suggested. Therefore, words such as *warning*, *fear* and *fearful*, *worries* and *deadly* are unique to the popular representation of the topic. This is in line with the use of conceptual metaphors that refer to war and also to the 9/11 attacks. In transferring those attributes (*war*, *victims*, *terror*) to the epidemic, specific aspects of SARS are related to current public concerns and anxieties surrounding the coverage after the U.S.A. proclaimed their "war on terror".

Those metaphors are also often used in combination with vague quantifiers to express pensiveness (*SARS claims many victims*) and, in this context, Bell's (1996: 216) statement that "the findings of science are often in news terms almost qualified to death" and are therefore altered seems quite appropriate. So, from a functional perspective, the difference in style is closely connected with the rhetorical aims and the preference for vague quantifiers over concrete numbers in the popular discourse is an expression of that. The corpus statistics in section 4.2 have shown that there is a significantly higher use of figures the medical texts and the prominent number of comparative adjectives and adverbs in the popular subcorpus, i.e. relational quantifiers such as *more*, *some*, *many*, *most*, *other* and *all*, is the mirror-image story. The example of the contextualisation of *patient(s)*, *case(s)* and *people* has also given evidence for the process of priming since there is a systematic co-selection of particular figures and quantifiers with these words.

The answer to the question as to if there is a different lexical realisation of a scientific topic in the two genres and the answer must be 'yes, to some extent' as there are also lexical patterns and structures that correlate. From a lexical perspective, the two discourse forms are not divided by a vast gulf, but by a gulf that can be bridged. Science and public meet halfway and are intertwined as they share a common stock of words. There is, however, a divergent preference in lexical choice that marks them as 'specific' or 'general'.

6.3 Towards a theory of popularisation

The overall research question stated in the introduction to this study was if there is a relation between scientific and popular discourse and if there is one, how it is shaped by the interplay between science and the media. As far as the coverage of a scientific topic in the newspress is concerned, a process of popularisation has to follow the rules of the genre since the construction of a topic is largely dependent on the genre routines and on the information that is generated from the situational context. The results of the analysis have made it clear that information about the disease was largely derived from its official sources. This is reflected by the focus on a certain element of the superstructure that was proposed by van Dijk (1988). Popular articles rely heavily on the *episodes* and the *comments* that are linked by the employment of expert opinion. In this connection, the use of epistemic modals has shown that verbal reactions are often used to place journalists' own assumptions and evaluations on reliable ground. The lexical realisation of a scientific topic in the press is always tied to the newsworthiness, the purpose and the motivation of what is said. Newsworthiness depends on the expectations and needs of the public audience which are certainly different from those of a 'scientific' audience. For the general public, the value of a scientific discussion is its actual outcome, not the way and methodology that lead to it. As Calsamiglia (2003: 14) says, "what is important is its application, its utility and the consequences of its use in people's lives". The corpus analysis has confirmed that the lexical realisation of public SARS discourse is mainly determined by two – almost conflicting – intentions:

- to inform and guide behaviour (e.g. to explain the spread of SARS = the realisation and contextualisation of *nasopharyngeal aspirates*);
- to persuade and entertain the reader (e.g. to capture the attention = the use of the conceptual 'war' metaphor = the realisation of *victims*).

In general, the comparison of the two collections of texts has shown that the popular texts focus more on the core of science rather than its details (selection and relevance) and that lexical choice reflects the tendency to employ items that are as familiar and vivid as possible (reduction of terminology). If the particular discourse segment is governed by the first intention (to inform, explain and guide behaviour), a process of lexical popularisation can be observed. If a scientific term is used, and there is no institutionalised equivalent at hand, use is limited to one piece of terminology, the technical term is embedded in a general environment and the foregrounded item is introduced to the reader. If the term itself is not required in the text, the use of classical (Latin) terms is omitted and the use of words of vernacular origin is preferred (lexical substitution). Other foregrounded linguistic devices, such as abbreviations, are either introduced or paraphrased. If the segment of text is governed by the second intention, the information is adapted to the genre routines, but not necessarily popularised. Genre routines of the newsmedia include the emotional appeal (negative semantic prosody), use of official quotes and comments as well as the relation to relevance for everyday life (generalisation, vagueness). All in all, the lexical processes that can be qualified as popularisation of scientific discourse are:

- reduction of complexity (by omitting detailed scientific information, but elaborating on pieces of information, presenting in referential context);
- simplification of scientific concepts (substitution, use of institutionalised or general forms on the top level of a taxonomy);
- re-contextualisation of items (embedding in a general context, use of explanatory devices, adaptation to genre routines);
- illustration (metaphors, relate to the competence of the audience).

Corpus analysis has also highlighted the fact that not only medical but also general terms are used side by side in a specific way. In this context, the theory of priming has provided a valuable framework for the interpretation of contextual patterns. The scientific and the 'public' use of items is bound to specific norms and routines (e.g. the realisation of the concept *affected persons*). This influences not only the choice of particular words but also their collocational behaviour, semantic prosody and idiomaticity. Research on scientific discourse popularisation should therefore not only focus on the reduction of scientific terms or the substitution of terminology but should also pay attention to rhetorical means and the affirmation of genre routines through priming. Corpus analysis has thus confirmed that some divergent patterns are due to the different constructions of the topic by the media. In analysing the popular presentations of the SARS crisis, the characteristics of popular scientific discourse coverage in the press must be sought in context and we have seen that this type of discourse adapts information from specific language and transfers them into its own preferred way of expression. Abstract information is related to everyday life by the use of conceptual metaphors, tentative language is used to hedge unwarranted claims, numerical information is expressed by the use of quantifiers and, most importantly, negative semantic prosody is applied to evoke an emotional appeal.

At this point, I would like to come back to the theoretical discussion outlined in Chapter 2. Although the line of study is a fairly recent one, there is already a *dominant view* of scientific discourse popularisation which suggests that there is a linear one-way simplification process from scientific to popular texts. This implies that there are two separate discourses, of which one is the basis for the other, and information is debased and transferred from scientific to popular texts in simplified form. Bell (1996), and also Gotti (2005), refer to it as a kind of translation process:

> Here the journalist acts as text processor, [...], but the technical subject matter makes her role of translator and interpreter more obvious. Medical terms are translated, conclusions picked up from the end. (Bell 1996: 217)

> Another aspect that suggests this analogy between popularisation and translation is the fact that both processes tend to produce an imperfect equivalence of a source text. A popularised text approximates the content of the original text in the same way that translation diverges from its source text. (Gotti 2005: 205)

As far as the scientific popular realisation of the SARS discourse is concerned, the corpus-based study has provided evidence that this view has to be questioned. There is obviously a point of interception between the two collections of texts and they are clearly intertwined as they share a common stock of words. However, form and contextualisation of shared concepts can differ. Differences in the individual realisation are more than the result of a popularisation process; they are apparently tied to essential and unshiftable characteristics and routines of respective genre. Those who also question the dominant view, on the other hand, claim that the popular report of a scientific topic in the general press is not related to its specific counterpart at all and that it would be wrong to see a causal relation between the two. Gledhill sees them as "simply stories in their own right, with their own phraseology and which belong to different discourses" (Gledhill 1996: 3).

7 Summary and prospects for future research

The main objective of the present study has been to contribute a model of lexical processes to the discussion of possible mechanisms in scientific discourse popularisation. To this end, the theoretical framework of special language and its division was outlined and a new approach to the verticality of popularisation was developed in section 2. Section 3 illustrated the methodological steps that were taken towards the construction of such a model and described the compilation and design of the specialised corpus that was used for the large-scale computational analyses. The study of word frequencies in section 4 shed some empirical light on the lexical choice in the two collections of texts and a categorisation was established that pictures this relation quantitatively. Prominent words in terms of frequency were then subject to detailed lexical profiling that took the contextualisation of general words, institutionalised terms as well as of specific terminology into account (section 5). This provided the basis for the description of shared patterns and individual traits in section 6, which led to the establishment of a lexical model that sets the theoretical framework of scientific discourse as well as contextual and co-textual features in interrelation. On the way towards a description of the lexis used at the interface of science and the general public, the study has raised some interesting points that provide valuable insights:

- for the study of Language for Specific Purposes,
- for a theory of Scientific Discourse Popularisation,
- for aspects of lexicology and lexical semantics, and
- for methodological issued in corpus-linguistics.

For decades, the focus of LSP research has been on the horizontal division of special languages but now a shift towards the vertical dimension has taken place as scientific topics are no longer discussed within the ivy-clad walls of research circles only. For linguists, it is certainly of interest to observe how a particular discourse that steps out of its special language limits is constructed and how genre routines are affirmed. From a lexical point of view, science is mirrored in everyday language use but this mirror shows an image that has been reshaped. Deviations in the individual realisation of shared topics apparently result from essential and unshiftable characteristics of the genre to what it is transferred. The representation of SARS in the general newspress has shown that popularisation is not about the transfer of information only but it is even more so about the realisation of genre routines, such as the use of conceptual metaphors. However, as far as the debate about a possible connection between science and society is concerned, the present study has proved that there is a point of intersection between the two worlds and they should not to be regarded as divided by a vast gulf.

From a lexical perspective, the respective choice proves that general language items, institutionalised terms and specific terminology are used side by side in both genres. The two collections of texts share a common stock of words on a high frequency basis, however, the identical form of these words does not automatically lead to an identical contextualisation and usage. Identical forms might function differently (e.g. *virus*), whereas words that are not identical in form are related in meaning and function (e.g. *HCWs*, *doctors and nurse*). Most interestingly, the lexical profiling for the candidate patterns has revealed certain processes of recontextualisation that can be interpreted in terms of popularisation, functional dialect or lexical priming. Hoey (2005: 12) claims that "priming is the driving force behind language use, language structure and language change". This view holds that a word is placed into a context that already exists but Hoey turns this process around and claims that the selection of an item triggers the actual context. Evidence for this is the (re-)contextualisation of numerical information that regularly collocates with words like *patients*, *cases* or *people*. In the medical text, *patients* is primed for use with precise figures at a moderate level (e.g. *3 patients*), whereas *patients* is in the popular texts primed to collocate with vague quantifiers, such as *many* or *some*. So, lexical priming might indeed contribute to a theory of popularisation as it has an individual nature and is highly genre- and domain-specific. The same is true of the concept of semantic prosody. The realisation of shared words and concepts can be seen as semantically distinct in terms of negative connotations. Words used to express and evoke negative emotive states, such as *fear* or *worries*, were found to occur in the popular texts only. This has not so much do with priming than with the individual genre routines. While emotional appeal is a device that is usually employed in the construction of news stories, the very absence of emotive language is a characteristic in scientific discourse. Most interestingly, institutionalised words, such as *virus* that have usually a neutral prosody, take on certain 'popular' semantic prosodies when they are used outside of their source domain (e.g. *killer virus*). Nevertheless, the two collections of texts display some similarities in relation to preference for certain semantic sets although, in the medical texts, there is a distinct tendency to narrow and specify the meaning of the node by the co-selection of given semantic sets. This is true both of scientific terms as well as of general words.

As the use of corpora in LSP research has been expanded just within the last decade and the study of external communication is a fairly recent trend, this study sheds light on a new perspective by applying corpus-linguistic methods to the study of scientific discourse popularisation. Nevertheless, as Partington

(1998) states, such a corpus-linguistic approach has its shortcomings, which have to be taken into account:

> A corpus, no matter how large and varied, is only ever representative of itself, and claims made about the behaviour of itself and claims that are made about the behaviour of linguistic items after studying corpus data should bear this in mind. (Partington 1998: 146)

It goes without saying that the results of this study can only provide a snapshot of the relation between the worlds of science and the general public; however, the corpus data has proved to provide an effective basis for the observation of shared patterns and individual traits. I would also argue that the use of corpora for the investigation of science popularisation has a considerable advantage: it is not based on intuition. Intuitive selection of candidate patterns in the texts might well trigger the very items that are foregrounded as they have an immediate eye-catching effect. The assumptions drawn from such an analysis might neglect the (not foregrounded) common ground that actually exists between the two discourse forms. With the computer-assisted corpus analysis, words are initially viewed from a frequency-based perspective and the selection of items is based on their prominence in terms of quantitative standing.

The research question demanded a fairly specialized corpus and the decision to compile such a comparable corpus myself has been proved to be reasonable. The comparability of the two subcorpora, in terms of topic, distribution and time frame, has provided an exceptional opportunity to study related patterns as well as individual realisations that are due to the different genre routines and the particular audience design. The pilot study that preceded the large-scale analysis was also a step worth taking as it put the corpus design and the methodology to an initial test. It has proven that there should also be room for small-scale studies as they point the way in which research should be directed. Methodologically, the combination of basic-wordlist, refined-wordlist and key-wordlist analysis turned out to be most effective for a comprehensive overview of lexical choice in the two subcorpora. I would argue that without this multiple approach to frequencies, the scaffolding of quantitative categorisation that can be provided for a qualitative analysis would not have been so useful. The same holds true for the combination of the interpretation of concordance lines by applying *hypothesis testing* with the analysis of the most frequent collocates and clusters in Chapter 5.

I would thus encourage the use of corpora in scientific discourse popularisation and make a plea for extending the model that I have proposed. This study focuses on lexis and is thus merely only a first contribution and a fragment of what should become a comprehensive model of scientific discourse popularisation. Ideally, it should include a syntactical, pragmatic, textual and sociolinguistic view on the relation between science and society. Furthermore, studies of popularisation and knowledge transfer have also pointed out the need of discussing popularisation at different levels on the vertical scale (cf. Niederhauser 1999, Wichter and Antos 2001). This is what I have referred to as "verticality within the verticality". The different mode of popularisation is immediately dependent on the intention behind the process, i.e. educating, informing, entertaining and so on. In addition, the study of scientific discourse popularisation has to step out of certain domain limits in order to depict general patterns across the borders of specific terminologies. Although the topic that I have analysed in the present study is a limited one, I would like to emphasise the relevance of the phenomenon of global epidemics as a recurrent one in the mass media. In today's worldwide mobility, emerging diseases, such as *SARS* or *bird flu*, sometimes reach pandemic proportions and thus have an effect on the public at large. In this connection, a diachronic view on the early popularisation of diseases would also be promising. Here, it would be useful to stay within a certain field in order to investigate how, for example, information about *the plague* was distributed among the public in medieval times. Such a study could be corpus-based as, for example, a corpus of historical texts could be compared to a corpus of early medical writing. Popularisation could be thus traced through the ages.

The study of scientific discourse popularisation is all set to have a bright future. I believe that Myers (2004) is right in criticising the dominant view that has already been established in this emerging branch of LSP research. I join him in questioning the boundaries of scientific discourse popularisation in the newspress. However, I also make a plea for not regarding the different realisations as a vast gulf dividing science and society but to regard it as a bridge between two worlds. If we take all similarities and differences into account, it becomes clear that popular discourse is not a simplified, derived version of a scientific source text. The two genres are related to each other in various ways but take account of the different routines and the individual competences of their target audience.

References

Aarts, J. (2002): "Does corpus linguistics exist? Some old and new issues", Breivik, L.E. (ed.): *From the COLT's Mouth ... and Others': Language Corpora Studies in Honour of Anna-Brita Stenström*. Amsterdam: Rodopi.

Adams-Smith, D. (1987): "Variation in field related genres", *Genre Analysis and ESP,* Vol. 1, 10-32.

Adler, R. and Rodman, G. (2000): *Understanding communication*. Fort Worth: Harcourt.

Antos, G. and Wichter, S. (eds.) (2001): *Wissenstransfer zwischen Experten und Laien. Umriss einer Transferwissenschaft.* Frankfurt/Main: Peter Lang.

Aston, G. (2001): *Learning with corpora: an overview.* Bologna: CLUEB.

Atkinson, D. (1992): "The Evolution of Medical Research Writing from 1735 to 1985: The Case of the Edinburgh Medical Journal", *Applied Linguistics* 13 (4), 337-374.

Barnbrook, G. (1996): *Language and Computers: A Practical Introduction to the Computer Analysis of Language.* Edinburgh: Edinburgh University Press.

Baumann, K.-D. and Kalverkämper, H. (eds.) (1992): *Kontrastive Fachsprachenforschung.* Tübingen: Narr.

Baumann, K.-D. (1992): *Integrative Fachtextlinguistik.* Tübingen: Narr.

Baumann, K.-D. (2001): "Cognitive turn in LPS research", Mayer, F. (ed.): *Language for Special Purposes: Perspectives for the New Millennium* Vol. 1, 87-102.

Becker, A. (2001): *Populärmedizinische Vermittlungstexte. Studien zur Geschichte und Gegenwart fachexterner Vermittlungsvarietäten.* Tübingen: Niemeyer.

Beier, R. (1980): *Englische Fachsprache.* Stuttgart: Kohlhammer.

Berber Sardinha, T. (1999): "Word sets, keywords, and text contents - an investigation of text topic on the computer", *DELTA* 15(1), 141-149.

Bell, A. (1984): "Language style as audience design. *Language in Society* 1(3), 145-204.

Bell, A.(1996): *The Language of the News Media.* Oxford: Blackwell

Bell, A. (2001): "Back in style: reworking audience design", Eckert, P. and Rickford, J. (eds.): *Style and Sociolinguistic Variation.* New York: Cambridge University Press.

Beneš, E. (1969): "Zur Typologie der Stilgattungen in wissenschaftlicher Prosa", *Deutsch als Fremdsprache* Vol. 3, 227.

Benson, M., Benson, E. and Ilson, R. (1986): *The BBI Combinatory Dictionary of English: A Guide to Word Combinations.* Amsterdam: John Benjamins.

Biber, D. (1988): *Variation Across Speech and Writing*. Cambridge: Cambridge University Press.

Biber, D. (1993): "Representativeness in Corpus Design", *Literary and Linguistic Computing* 8(4), 243-257.

Biber, D., Conrad, S. and Reppen, R. (1998): *Corpus Linguistics. Investigating language structure and use.* Cambridge: Cambridge University Press.

Biber, D. (1999): "A register perspective on grammar and discourse: variability in the form and use of English complement clauses", *Discourse Studies* 1 (2), 131-150.

Biere, B. U. and Liebert, W.-A. (1997): *Metaphern, Medien, Wissenschaft. Zur Vermittlung der AIDS-Forschung in Presse und Rundfunk.* Opladen: Westdeutscher Verlag.

Bloomfield, L. (1939): *Linguistic Aspects of Science.* International Encyclopedia of Unified Science. Vol. 1(4). Chicago: University of Chicago Press.

Bowker, L. and Pearson, J. (2002): *Working with Specialized Language: a Practical Guide to Using Corpora.* London: Routledge.

Brand, C. (2002): *Overcoming the Linguistic Barrier.* Unpublished diploma thesis. University of Giessen.

Brinton, L. (2000): *The Structure of Modern English: a Linguistic Introduction.* Amsterdam: John Benjamins.

Buntgarten, T. (ed.) (1993): *Beiträge zur Wirtschaftskommunikation*. Tostedt: Attikon.

Busch, A. (1994): *Laienkommunikation. Vertikalitätsuntersuchungen zu medizinischen Experten-Laien-Kommunikationen*. Frankfurt am Main: Peter Lang.

Busch, A. (2001): "Wissenstransfer und Kommunikation in Gesundheitszirkeln", Antos, G. and Wichter, S. (eds.): *Wissenstransfer zwischen Experten und Laien: Umriss einer Transferwissenschaft.*

Calsamiglia, H. (2003): "Popularization Discourse", *Discourse Studies* 5(2), 139-146.

Calsamiglia, H. and Lopez-Ferrero, C. (2003): "Role and position of scientific voices: reported speech in the media", *Discourse Studies* 5(2), 147-173.

Canale, M. and Swain, M. (1980): "Theoretical bases of communicative approaches to second language teaching and testing", *Applied Linguistics*, 1(1), 8-24.

Candel, D. (1994); "Le discours definitoire: variations discursives chez les scientifiques", Moirand, S. et al. (eds.): *Parcours Linguistiques de Discours Specialises*, 33–34.

Carter, R. (1987): *Vocabulary: Applied Linguistics Perspectives*. London: Longman.

Chargaff, E. and Allen, U. (1986): "How scientific papers are written", *Fachsprache* 8(3-4), 106-111.

Charteris-Black, J. (2005): *Politicians and Rhetoric: the Persuasive Power of Metaphor*. Houndmills: Palgrave Macmillan.

Ciaspusco, G. E. (2003): "Formulation and reformulation procedures in verbal interactions between experts and (semi-)laypersons", *Discourse Studies* 5 (2), 207–233.

Cohen, R.(1986): "On the relationship between user models and discourse models", *Computational Linguistics* 14(3), 88-90.

Collins COBUILD English Language Dictionary (1991), 5th Edition., London : HarperCollins.

Cooter, R. and Pumfrey, S. (1994): "Separate Spheres and Public Places: Reflections on the History of Science Popularisation and Science in Popular Culture", *History of Science* Vol. 32, 237-267.

Crombie, W. (1985): *Discourse and Language Learning: a Relational Approach to Syllabus Design*. Oxford: Oxford University Press.

Crystal, D. (1999): *The Cambridge Encyclopaedia of Language*. Cambridge: Cambridge University Press.

Crystal, D. (2003): *The Cambridge Encyclopaedia of the English Language*. Cambridge: Cambridge University Press.

Daneš, F. (1964): "A three-level approach to syntax", *Travaux Linguistiques de Prague 1*, 225-240.

De Beaugrande, R. and Dressler, W. (1981): *Introduction to Text Linguistics*. London: Longman.

Dirckx, J. (1988): *The Language of Medicine: its Evolution, Structure and Dynamics*. New York, NY: Praeger.

Dorlands Medical Dictionary (2003), 30th Edition, New York: Harcourt.

Drodz, L. and Seibicke, W. (1973): *Deutsche Fach- und Wissenschaftssprache. Bestandsaufnahme – Theorie – Geschichte*. Wiesbaden: Brandstetter.

Dunning, T. (1993): "Accurate methods for the statistics of surprise and coincidence", *Computational Linguistics* 19(1), 61-74.

Eckert, P. and Rickford, J. (eds.) (2001): *Style and Sociolinguistic Variation*. New York and Cambridge: Cambridge University Press.

Esser, J. (1993): *English Linguistic Stylistics*. Tübingen: Niemeyer.

Esser, J. (1999): "Collocation, colligation, semantic preference and semantic prosody: new developments in the study of syntagmatic word relations", Falkner, W. and Schmid, H.-J. (eds.): *Kolloquium für Leonhard Lipka*. Tübingen: Narr.

Firth, J. (1957): *Papers in Linguistics*. London: Oxford University Press

Fischer, R. (1998): *Lexical change in present-day English: a corpus based study of the motivation, institutionalization, and productivity of creative neologisms*. Tübingen: Narr.

Fluck, H.-R. (1991): *Fachsprachen*. Tübingen: Francke.

Francis, W.N. (1992): "Language corpora B.C.", Svartvik, J. (ed.): *Directions in Corpus Linguistics: Proceedings of Nobel Symposium 82*. Berlin: Mouton de Gruyter, 17-32.

Garvin, P. L. (ed.): *A Prague School Reader on Esthetics, Literary Structure, and Style*. Washington: Washington University Press.

Gerzymisch-Arbogast, H. (1987): *Zur Thema-Rhema-Gliederung in amerikanischen Wirtschaftsfachtexten. Eine exemplarische Analyse*. Tübingen: Narr.

Giles, H. (1973). "Accent mobility: A model and some data", *Anthropological Linguistics* Vol. 15, 87-105.

Giles, H., Taylor, D.M., and Bourhis, R. (1973): "Towards a theory of interpersonal accommodation through language: Some Canadian data", *Language in Society* Vol. 2, 177-192.

Gläser, R. (1979): *Fachstile des Englischen*. Leipzig: VEB Verlag.

Gläser, R. (1990): *Fachtextsorten im Englischen*. Tübingen: Narr.

Gläser, R. (1992): "A multi-level model for a typology of LSP genres", *Fachsprache* 15(1-2), 18-26.

Gläser, R. (1995): *Linguistic Features and Genre Profiles*. Frankfurt/M: Peter Lang.

Gledhill C. (1996): "The phraseology of rhetoric, collocations and discourse in cancer research abstracts", Barron, C. and Bruce, N. (eds.): *Knowledge and Discourse: Proceedings of the International Multidisciplinary Conference*. Hong Kong: University of Hong Kong.

Gledhill, C. (2000): *Collocations in science writing*. Tübingen: Narr.

Gotti, M. (2005): *Investigating specialized discourse*. Bern: Peter Lang.

Gülich, E. (2003): "Conversational techniques used in transferring knowledge between medical experts and non-experts", *Discourse Studies* 5(2), 235-263.

Gülich, E. and Kotschi, T. (1987): "Reformulierungshandlungen als Mittel der Textkonstitution. Untersuchungen zu französischen Texten aus mündlicher Kommunikation", Motsch, W. (ed.): *Satz, Text, sprachliche Handlung*. Berlin: Akademie Verlag.

Haage, B. (1993): "Wissenstradierende und gesellschaftliche Konstituenten mittelalterlicher deutscher Fachsprache", Buntgarten, T. (ed.): *Beiträge zur Wirtschaftskommunikation*.

Havránek, B. (1964 (orig. 1932)): "The functional differentiation of the standard language", Garvin, P. L. (ed.): *A Prague School Reader on Esthetics, Literary Structure, and Style*.

Helbig, G. (1986): *Geschichte der neueren Sprachwissenschaft*. Opladen: Westdeutscher Verlag.

Highfield, R. (2000): "Essays on science and society: Selling Science to the Public", *Science* 289(5476), 59.

Hilgartner, S. (1990): "The dominant view of popularization", *Social Studies of Science* 20, 519–39.

Hill, J. (1999): "Collocational competence", *English Teaching Professional* Vol. 11, 3-6.

Hoffmann, L. (1976): *Kommunikationsmittel Fachsprache*. Eine Einführung. Tübingen: Narr.

Hoffmann, L. (1985): *Vom Fachwort zum Fachtext*. Tübingen: Narr.

Hoffmann, L. (1990): *Fachtexte und Fachtextsorten*. Leipzig: Karl-Marx-Universität.

Hoffmann, L., Kalverkämper, H. and Wiegand, H. E. (1998): *Fachsprachen. Ein internationales Handbuch zur Fachsprachenforschung und Terminologiewissenschaft*. Berlin: de Gruyter.

Hoey, M. (1991): *Patterns of Lexis in Text*. Oxford University Press.

Hoey, M. (1997): "From concordance to text structure: new uses for computer corpora", Melia, J. and Lewandoska, B. (eds.): *PALC 97: Practical Applications in Language Corpora,* Lodz, Poland: Lodz University Press.

Hoey, M. (2005): *Lexical Priming: a New Theory of Language*. London: Routledge.

Hunston, S. (2002): *Corpora in Applied Linguistics*. Cambridge: Cambridge University Press.

Hyland, M. (1998): *Hedging in Scientific Research Articles*. Amsterdam: John Benjamins.

Hymes, D. (1972) "On communicative competence", Pride, J. and Holmes, J. (eds.) *Sociolinguistics,* 269-293.

Jakob, K. (1998): "Fachsprachliche Phänomene in der Alltagskommunikation", *HSK* 1, 710-717.

Jakobson, R. (1963): "Implications of language universals for linguistics", Greenberg, J. H. (ed.): *Universals of Language*. Cambridge, Mass.: MIT Press.

Johns, T. (1991): "Should you be persuaded - two examples of data-driven learning materials", *English Language Research Journal* Vol. 4, 1-16.

Johnson, E. (1994): "Police speak", Brekke, M. (ed.): *Applications and implications of current LSP research: proceedings of the 9th European Symposium on LSP, Aug. 2-6, 1993*. Bergen: Fagbokforlaget.

Kalverkämper, H. (1988): "Die Fachwelt in der allgemeinen einsprachigen Lexikographie" *Fachsprache* (10), 97-122.

Karpf, A. (1988): *Doctoring the Media: The Reporting of Health and Medicine*. London: Routledge.

Kjær, A. L. (1990): "Phraseology research – state-of-the-art", *Terminology Science & Research, Journal of the International Institute for Terminology Research* – IITF 1: 3-20.

Kennedy, G. (1998): *An Introduction to Corpus Linguistics*. London, New York: Longman.
Lakoff, G. and Johnson, M. (1980): *Metaphors we Live by*. Chicago: University of Chicago Press.
Laurén, C. (1994): "LSP - a valid concept in 1993?", Brekke, M. (ed.): *Applications and Implications of Current LSP Research*. Bergen: Fakbokforlaget.
Lee, D. (2001): *Cognitive Linguistics: an Introduction*. South Melbourne, Vic.: Oxford University Press.
Lee, D. (2001): "Genres, registers, text types; domains, and styles", *Language Learning & Technology* Vol. 5(3), 37-72.
Leech, G. (1974): *Semantics*. Harmondworth: Penguin.
Leech, G. (1991): "The state of the art in corpus linguistics", *English Corpus Linguistics: Linguistic Studies in Honour of Jan Svartvik*. London: Longman, 8-29.
Leech, G. (2001): *Word Frequencies in Written and Spoken English*. London: Longman.
Lipka, L. (1992): *An outline of English lexicology :Lexical Structure, Word Semantics, and Word-Formation*. Tübingen: Niemeyer.
Logan, R. A. (2001): "Science mass communication", *Science Communication* 23(2), 135-163.
Louw, B. (1993): "Irony in the text or insincerity in the writer? The diagnostic potential of semantic prosodies", Baker, M., Francis, G. and Tognini-Bonelli, E. (eds.) *Text and Technology*. In Honour of John Sinclair. Philadelphia: John Benjamins.
Lyons, J. (1981): *Language and Linguistics*. Cambridge: Cambridge University Press.
Macdonald, K. M. (1995): *The Sociology of the Professions*. Thousand Oaks, Calif.: Sage.
Macmillan English Dictionary for Advanced Learners (2002), Oxford: Macmillan.
Mathesius, V. (1936/1964): "On some problems of the systematic analysis of Grammar", *Travaux linguistiques de Prague* 6: 95-107. Reprinted in Vachek (ed.) 1964.
Mathews, H. (1991): *Morphology*. Cambridge: Cambridge University Press.
Mayer, F. (ed.) (2001): *Language for special purposes: Perspectives for the new Millennium. Linguistics and Cognitive Aspects, Knowledge Representation and Computational Linguistics, Terminology, Lexicography and Didactics.* Tübingen: Narr.
McEnery, T. and Wilson, A. (1996): *Corpus Linguistics*. Edinburgh: Edinburgh University Press.

Michler, M. (1981): *Einführung in die medizinsche Fachsprache.* Berlin: Springer.
Moirand, S. (ed.) (1994): *Parcours linguistiques de discours specialises.* Bern: Peter Lang.
Moirand, S. (1997): "Formes discursives de la diffusion des savoirs dans les medias", *Hermès, Cognition, Communication, Politique* (21), 33-44.
Mukherjee, J. (2002): "The scope of corpus evidence", Brend, R. M., Sullivan, W. J. and Lommel, A. R. (eds.): *LACUS Forum XXVIII: What Constitutes Evidence in Linguistics.* Houston, TX: The Linguistic Association of Canada and the United States, 103-114.
Mukherjee, J. (2004): "The state of the art in corpus linguistics: three book-length perspectives", *English Language and Linguistics* 8(1), 103- 119.
Mukherjee, J. (2005): "Word", Strazny, P. (ed.): *Encyclopedia of Linguistics*, New York: Fitzroy Dearborn, 1184-1186.
Myers, G. (2003): "Discourse studies of scientific popularization: questioning the boundaries", *Discourse Studies* Vol. 5(2), 265-279.
Nation, P., and Waring, R. (1997): "Vocabulary size, text coverage, and word lists", Schmitt, N. and McCarthy, M. (eds.): *Vocabulary: Description, Acquisition, Pedagogy.* New York: Cambridge University Press.
Nelson, M. (2000): *A Corpus-Based Study of Business English and Business English Teaching Materials.* Unpublished PhD Thesis. Manchester: University of Manchester. Available at: http://users.utu.fi/micnel/thesis.html. Accessed at 9. January 2005.
Nelson, M. (2006): "Semantic associations in Business English. A corpus-based analysis", *English for Specific Purposes* 25, 217-234.
Nesselhauf, N. (2005): *Collocations in a Learner Corpus.* Amsterdam: Benjamins.
Newmeyer, F. J. (2000): *Language Form and Language Function.* Cambridge, MA: MIT Press.
Niederhauser, J. (1999): *Wissenschaftssprache und Populärwissenschaftliche Vermittlung.* Tübingen: Narr.
Nwogu, K.N. (1991): "Structure of science popularizations: genre-analysis approach to the schema of popularized medical texts", *English for Specific Purposes* 10(2), 111-123.
Oakman, R.L. (1980): *Computer Methods for Literary Research.* Columbia: University of South Carolina Press.
Oakes, M.P. (1998): *Statistics for Corpus Linguistics.* Edinburgh: Edinburgh University Press.
Palmer, F. (1990): *Modality and the English modals.* London: Longman.
Partington, A. (1998): *Patterns and Meanings: Using Corpora for English Language Research and Teaching.* Amsterdam: Benjamins.

Partington, A. (2004): "Utterly content in each other's company. Semantic prosody and semantic preference", *International Journal of Corpus Linguistics* 9(1), 131-156.

Paolillo, J. C. (2002): *Analyzing Linguistic Variation: Statistical Models and Methods*. Stanford: CSLI Publications.

Pride, J.B. and Holmes, J. (eds.) (1985): *Sociolinguistics*. Harmondsworth: Penguin.

Pschyrembel (2004), 260th Edition, Berlin: Mouton de Gruyter.

Radford, T. (2001): "Language – the barrier between science and the public", *Croatian Medical Journal* 42(4), 353-355.

Rijn van Tongeren, G. (1997): *Metaphors in medical texts*. Amsterdam: Rodopi.

Roelcke, T. (2005): *Fachsprachen*. Berlin: Schmidt.

Rothbart, D. (1997): *Explaining the Growth of Scientific Knowledge: Metaphors, Models and Meanings in Science*. New York: Edwin Mellen Press.

Russell, C. (1986): "The view from the national beat", Friedman, S. M. (ed.): *Scientists and Journalists: Reporting Science as News*. New York: Free Press.

Sager et al. (1980): *English Special Languages: Principles and Practice in Science and Technology*. Wiesbaden: Brandstetter.

Schröder, H. (ed.) (1991): *Subject-Oriented Texts. Languages for Special Purposes and Text Theory*. Berlin, New York: de Gruyter.

Scott, M. (2004): *WordSmith Tools*, Ver. 4.0, OUP. Available online at: http://www.lexically.net/wordsmith/index.html.

Scott, M. (2004): *WordSmith Tools Manual*. Ver. 4.0. Oxford: Oxford University Press.

Scott, M. and Tribble, C. (2005): *Working with Texts: Keywords and Keyword Analysis in Language Education*. Amsterdam: John Benjamins.

Sinclair, J. (1987): "The nature of the evidence", Sinclair, J. (ed.): *Looking up*. Glasgow: Collins.

Sinclair, J. (1991): *Corpus, Concordance, Collocation*. Oxford: Oxford University Press.

Sinclair, J. (1999): "A way with common words", Hasselgård, H. and Oksefjell, S. (eds.): *Out of Corpora*. Amsterdam: Rodopi.

Sinclair, J. (1996): "The search for units of meaning", *Textus*, Vol.9, 75-105.

Sinclair, J. (2004): *Trust the Text. Language, Corpus and Discourse*. London: Routledge.

Skudlik, S. (1990): *Sprachen in den Wissenschaften. Deutsch und Englisch in der Internationalen Kommunikation*. Tübingen: Narr.

Steger, H. (1988): "Erscheinungsformen der deutschen Sprache. Alltagssprache - Fachsprache - Standardsprache - Dialekt und andere Gliederungstermini", *Deutsche Sprache* 16(88), 289-319.

Stichweh, R. (1984): *Zur Entstehung des Modernen Systems Wissenschaftlicher Disziplinen*. Frankfurt/M: Suhrkamp.
Stubbs, M. (1995): "Corpus evidence for norms of lexical collocation". Cook, G. and Seidlhofer, B. (eds): *Principle & Practice in Applied Linguistics. Studies in Honour of H.G. Widdowson*. Oxford: Oxford University Press.
Stubbs, M. (1996): *Text and Corpus Analysis: Computer Assisted Studies of Language and Institutions*. Oxford: Blackwell.
Swales, J. M. (1990): *Genre Analysis*. Cambridge: Cambridge University Press.
Tribble, C. (1998): "Genres, keywords, teaching: towards a pedagogic account of the language of project proposals", Burnard, L. (ed.): *Teaching and Language Corpora 98 Proceedings of the 1998 TALC conference*. Oxford: Oxford University Press.
Trimble, L. (1985): *English for Science and Technology: A Discourse Approach*. Cambridge: Cambridge University Press.
Tognini Bonelli, E. (2001): *Corpus Linguistics at Work*. Amsterdam: Benjamins.
Vachek, J. (ed.) (1964): *A Prague School Reader in Linguistics*. Bloomington: Indiana University Press.
van Dijk, T. A. (1988): *News as discourse*: Hillsdale, NJ: L. Erlbaum Associates.
Vestergard, T. and Schröder, K. (1985): *The Language of Advertising*. Oxford: Blackwell.
von Hahn, W. (1980): "Fachsprachen", *Lexikon der Germanistischen Linguistik*. Tübingen: Niemeyer.
Weigold (2001): "Communicating science", *Science Communication* Vol. 23(2), 164-193.
Wichter, S. (1994): *Experten- und Laienwortschätze. Umriß einer Lexikologie der Vertikalität.* Tübingen: Niemeyer.
Wichter, S. and Antos, G. (2001): *Wissenstransfer zwischen Experten und Laien. Umrisse einer Transferwissenschaft.* Frankfurt/M.: Peter Lang.
Wüster, E. (1931): *Internationale Sprachnormung in der Technik.* Berlin: VDI.
Xiao, Z. and McEnery, A. (2005): "Two approaches to genre analysis: three genres in modern American English", *Journal of English Linguistics* 33(1), 62-82.

Appendix: Wordlists

Basic wordlist for the popular subcorpus

N	*Word*	*Freq.*	*%*	*Texts*	*%*
1	THE	17.736	5,89	354	100,00
2	TO	8.183	2,72	352	99,44
3	OF	7.932	2,63	354	100,00
4	IN	6.888	2,29	353	99,72
5	A	6.333	2,10	353	99,72
6	AND	6.303	2,09	354	100,00
7	#	4.685	1,56	343	96,89
8	THAT	3.639	1,21	351	99,15
9	SARS	3.547	1,18	352	99,44
10	IS	2.945	0,98	341	96,33
11	S	2.502	0,83	334	94,35
12	IT	2.495	0,83	344	97,18
13	HAVE	2.347	0,78	340	96,05
14	FOR	2.322	0,77	333	94,07
15	SAID	2.098	0,70	299	84,46
16	WAS	1.868	0,62	315	88,98
17	WHO	1.863	0,62	314	88,70
18	FROM	1.850	0,61	329	92,94
19	ON	1.810	0,60	330	93,22
20	WITH	1.808	0,60	334	94,35
21	BE	1.762	0,59	323	91,24
22	HEALTH	1.752	0,58	322	90,96
23	ARE	1.725	0,57	305	86,16
24	AT	1.716	0,57	322	90,96
25	AS	1.675	0,56	321	90,68
26	HAS	1.642	0,55	328	92,66
27	DISEASE	1.543	0,51	303	85,59
28	BEEN	1.444	0,48	320	90,40
29	PEOPLE	1.400	0,46	311	87,85
30	BUT	1.339	0,44	318	89,83
31	VIRUS	1.312	0,44	278	78,53
32	HAD	1.299	0,43	296	83,62
33	BY	1.263	0,42	313	88,42
34	CASES	1.225	0,41	279	78,81

35	CHINA	1.206	0,40	276	77,97
36	HONG	1.178	0,39	272	76,84
37	KONG	1.169	0,39	270	76,27
38	THEY	1.159	0,38	278	78,53
39	NOT	1.139	0,38	294	83,05
40	WE	1.104	0,37	262	74,01
41	AN	1.082	0,36	298	84,18
42	WERE	1.064	0,35	273	77,12
43	HE	1.007	0,33	235	66,38
44	THIS	995	0,33	285	80,51
45	MORE	993	0,33	299	84,46
46	THERE	935	0,31	276	77,97
47	OR	934	0,31	276	77,97
48	NEW	900	0,30	277	78,25
49	OFFICIALS	835	0,28	225	63,56
50	HOSPITAL	808	0,27	201	56,78
51	ONE	773	0,26	247	69,77
52	WHICH	762	0,25	273	77,12
53	THAN	756	0,25	268	75,71
54	ABOUT	715	0,24	254	71,75
55	WORLD	684	0,23	304	85,88
56	THEIR	676	0,22	243	68,64
57	PATIENTS	674	0,22	203	57,34
58	ITS	668	0,22	251	70,90
59	WILL	659	0,22	221	62,43
60	NO	635	0,21	248	70,06
61	SPREAD	619	0,21	246	69,49
62	OTHER	616	0,20	240	67,80
63	T	606	0,20	186	52,54
64	INFECTED	591	0,20	226	63,84
65	MAY	591	0,20	235	66,38
66	IF	589	0,20	228	64,41
67	OUTBREAK	587	0,19	227	64,12
68	CAN	573	0,19	204	57,63
69	BEIJING	570	0,19	139	39,27
70	WOULD	567	0,19	224	63,28
71	COULD	562	0,19	231	65,25
72	I	558	0,19	135	38,14
73	GOVERNMENT	550	0,18	169	47,74
74	ALL	541	0,18	225	63,56
75	TWO	528	0,18	221	62,43
76	OUT	521	0,17	226	63,84

77	SO	520	0,17	213	60,17
78	FIRST	515	0,17	243	68,64
79	SOME	513	0,17	220	62,15
80	AFTER	511	0,17	229	64,69
81	WHEN	494	0,16	208	58,76
82	REPORTED	486	0,16	207	58,47
83	NOW	482	0,16	227	64,12
84	CHINESE	468	0,16	180	50,85
85	UP	466	0,15	212	59,89
86	LAST	465	0,15	199	56,21
87	HIS	460	0,15	142	40,11
88	PUBLIC	459	0,15	179	50,56
89	WHERE	458	0,15	222	62,71
90	ALSO	436	0,14	223	62,99
91	MOST	432	0,14	199	56,21
92	RESPIRATORY	432	0,14	263	74,29
93	SYMPTOMS	408	0,14	188	53,11
94	TORONTO	399	0,13	119	33,62
95	DOCTORS	397	0,13	150	42,37
96	MANY	394	0,13	183	51,69
97	BECAUSE	391	0,13	192	54,24
98	SHE	388	0,13	111	31,36
99	INTO	378	0,13	185	52,26
100	DAYS	375	0,12	192	54,24

Basic wordlist for the medical subcorpus

N	*Word*	*Freq.*	*%*	*Texts*	*%*
1	#	9.588	5,12	139	98,58
2	THE	8.700	4,64	141	100,00
3	OF	6.998	3,74	141	100,00
4	AND	5.538	2,96	141	100,00
5	IN	4.314	2,30	141	100,00
6	TO	3.783	2,02	139	98,58
7	SARS	3.001	1,60	137	97,16
8	A	2.999	1,60	140	99,29
9	WITH	2.281	1,22	136	96,45
10	FOR	1.882	1,00	136	96,45
11	PATIENTS	1.759	0,94	106	75,18
12	WERE	1.448	0,77	106	75,18
13	WAS	1.404	0,75	113	80,14
14	ON	1.312	0,70	132	93,62
15	THAT	1.269	0,68	133	94,33
16	IS	1.176	0,63	132	93,62
17	FROM	1.175	0,63	129	91,49
18	CASES	1.059	0,57	113	80,14
19	WHO	973	0,52	119	84,40
20	OR	971	0,52	110	78,01
21	HAVE	938	0,50	131	92,91
22	BY	924	0,49	123	87,23
23	AS	889	0,47	133	94,33
24	BE	807	0,43	126	89,36
25	HAD	795	0,42	101	71,63
26	HEALTH	769	0,41	111	78,72
27	AT	751	0,40	106	75,18
28	ARE	728	0,39	125	88,65
29	THIS	723	0,39	122	86,52
30	BEEN	721	0,38	132	93,62
31	RESPIRATORY	679	0,36	132	93,62
32	PATIENT	658	0,35	71	50,35
33	AN	606	0,32	117	82,98
34	VIRUS	594	0,32	91	64,54
35	HAS	584	0,31	127	90,07
36	NOT	566	0,30	117	82,98
37	CASE	552	0,29	101	71,63
38	CORONAVIRUS	550	0,29	87	61,70
39	HOSPITAL	538	0,29	76	53,90

40	CARE	536	0,29	72	51,06
41	WE	533	0,28	69	48,94
42	DISEASE	529	0,28	100	70,92
43	ALL	528	0,28	93	65,96
44	DAYS	514	0,27	85	60,28
45	OTHER	506	0,27	99	70,21
46	INFECTION	466	0,25	100	70,92
47	THESE	452	0,24	95	67,38
48	REPORTED	429	0,23	98	69,50
49	HONG	426	0,23	94	66,67
50	KONG	424	0,23	94	66,67
51	CLINICAL	379	0,20	76	53,90
52	AFTER	376	0,20	83	58,87
53	S	370	0,20	101	71,63
54	OUTBREAK	365	0,19	84	59,57
55	PROBABLE	365	0,19	73	51,77
56	SEVERE	343	0,18	128	90,78
57	TRANSMISSION	339	0,18	81	57,45
58	IT	331	0,18	96	68,09
59	ONE	330	0,18	94	66,67
60	CONTACT	324	0,17	76	53,90
61	COV	323	0,17	32	22,70
62	ACUTE	321	0,17	132	93,62
63	MORE	318	0,17	99	70,21
64	MAY	316	0,17	89	63,12
65	ALSO	315	0,17	95	67,38
66	NO	309	0,16	100	70,92
67	SYNDROME	307	0,16	131	92,91
68	BUT	306	0,16	97	68,79
69	OUR	304	0,16	56	39,72
70	SYMPTOMS	298	0,16	76	53,90
71	THERE	298	0,16	78	55,32
72	MARCH	296	0,16	59	41,84
73	THEIR	295	0,16	86	60,99
74	DURING	293	0,16	81	57,45
75	THAN	292	0,16	85	60,28
76	WHICH	289	0,15	107	75,89
77	CONTROL	281	0,15	88	62,41
78	ASSOCIATED	279	0,15	68	48,23
79	NEW	274	0,15	94	66,67
80	TWO	270	0,14	78	55,32
81	ILLNESS	261	0,14	66	46,81

82	CHINA	247	0,13	81	57,45
83	FEVER	237	0,13	53	37,59
84	VIRAL	233	0,12	50	35,46
85	SUSPECTED	227	0,12	55	39,01
86	TORONTO	225	0,12	53	37,59
87	HUMAN	220	0,12	50	35,46
88	INFECTED	217	0,12	75	53,19
89	TRAVEL	214	0,11	65	46,10
90	FIRST	213	0,11	78	55,32
91	THEY	210	0,11	85	60,28
92	USED	208	0,11	56	39,72
93	APRIL	207	0,11	54	38,30
94	PNEUMONIA	203	0,11	63	44,68
95	TIME	202	0,11	65	46,10
96	MOST	201	0,11	72	51,06
97	CHEST	199	0,11	41	29,08
98	SHOULD	199	0,11	65	46,10
99	IDENTIFIED	198	0,11	70	49,65
100	ONLY	197	0,11	82	58,16

Refined wordlist for the popular subcorpus

N	*Word*	*Freq.*	*%*	*Texts*	*%*
1	SARS	3.547	1,18	352	99,44
2	HEALTH	1.752	0,58	322	90,96
3	DISEASE	1.543	0,51	303	85,59
4	PEOPLE	1.400	0,46	311	87,85
5	VIRUS	1.312	0,44	278	78,53
6	CASES	1.225	0,41	279	78,81
7	CHINA	1.206	0,40	276	77,97
8	HONG	1.178	0,39	272	76,84
9	KONG	1.169	0,39	270	76,27
10	OFFICIALS	835	0,28	225	63,56
11	HOSPITAL	808	0,27	201	56,78
12	WORLD	684	0,23	304	85,88
13	PATIENTS	674	0,22	203	57,34
14	SPREAD	619	0,21	246	69,49
15	INFECTED	591	0,20	226	63,84
16	OUTBREAK	587	0,19	227	64,12
17	BEIJING	570	0,19	139	39,27
18	GOVERNMENT	550	0,18	169	47,74
19	REPORTED	486	0,16	207	58,47
20	CHINESE	468	0,16	180	50,85
21	LAST	465	0,15	199	56,21
22	PUBLIC	459	0,15	179	50,56
23	RESPIRATORY	432	0,14	263	74,29
24	SYMPTOMS	408	0,14	188	53,11
25	TORONTO	399	0,13	119	33,62
26	DOCTORS	397	0,13	150	42,37
27	BECAUSE	391	0,13	192	54,24
28	DAYS	375	0,12	192	54,24
29	WEEK	374	0,12	185	52,26
30	EPIDEMIC	373	0,12	154	43,50
31	HOW	370	0,12	154	43,50
32	TRAVEL	369	0,12	163	46,05
33	SINGAPORE	368	0,12	163	46,05
34	CITY	362	0,12	145	40,96
35	SAYS	362	0,12	95	26,84
36	MEDICAL	346	0,11	168	47,46
37	CONTROL	345	0,11	200	56,50
38	DEATHS	339	0,11	163	46,05
39	YESTERDAY	336	0,11	145	40,96

40	YEAR	327	0,11	171	48,31
41	GUANGDONG	326	0,11	141	39,83
42	STILL	320	0,11	170	48,02
43	SUSPECTED	317	0,11	148	41,81
44	CASE	312	0,10	157	44,35
45	DIED	308	0,10	158	44,63
46	SEVERE	306	0,10	249	70,34
47	NUMBER	305	0,10	165	46,61
48	THREE	301	0,10	168	47,46
49	WORKERS	294	0,10	139	39,27
50	JUST	292	0,10	153	43,22
51	HOSPITALS	291	0,10	123	34,75
52	TODAY	291	0,10	128	36,16
53	COUNTRIES	287	0,10	158	44,63
54	INFECTION	285	0,09	159	44,92
55	SCIENTISTS	285	0,09	112	31,64
56	ASIA	283	0,09	150	42,37
57	CANADA	279	0,09	134	37,85
58	ILLNESS	279	0,09	146	41,24
59	DAY	277	0,09	147	41,53
60	SYNDROME	276	0,09	242	68,36
61	LEAST	275	0,09	158	44,63
62	QUARANTINE	272	0,09	135	38,14
63	PROVINCE	270	0,09	152	42,94
64	ACUTE	261	0,09	241	68,08
65	FAR	260	0,09	153	43,22
66	HOME	258	0,09	131	37,01
67	CONTACT	256	0,09	151	42,66
68	COUNTRY	254	0,08	140	39,55
69	AUTHORITIES	253	0,08	143	40,40
70	ANOTHER	251	0,08	134	37,85
71	FLU	249	0,08	114	32,20
72	AGAINST	245	0,08	142	40,11
73	DEATH	242	0,08	126	35,59
74	PNEUMONIA	242	0,08	126	35,59
75	THROUGH	240	0,08	129	36,44
76	WHILE	240	0,08	145	40,96
77	AROUND	238	0,08	159	44,92
78	UNDER	237	0,08	167	47,18
79	DISEASES	234	0,08	124	35,03
80	CARE	231	0,08	96	27,12
81	INFECTIOUS	225	0,07	123	34,75

82	MIGHT	225	0,07	136	38,42
83	PATIENT	223	0,07	104	29,38
84	GLOBAL	218	0,07	118	33,33
85	INCLUDING	218	0,07	142	40,11
86	INFORMATION	216	0,07	111	31,36
87	US	216	0,07	120	33,90
88	UNIVERSITY	214	0,07	118	33,33
89	DR	212	0,07	98	27,68
90	SINCE	212	0,07	133	37,57
91	TAIWAN	211	0,07	85	24,01
92	MARCH	210	0,07	91	25,71
93	DOWN	209	0,07	131	37,01
94	WEEKS	207	0,07	134	37,85
95	MASKS	203	0,07	113	31,92
96	KNOWN	202	0,07	111	31,36
97	TOLD	199	0,07	120	33,90
98	SAY	198	0,07	117	33,05
99	CORONAVIRUS	195	0,06	73	20,62
100	DIRECTOR	194	0,06	130	36,72

Refined wordlist for the medical subcorpus

N	*Word*	*Freq.*	*%*	*Texts*	*%*
1	SARS	3.001	1,60	137	97,16
2	PATIENTS	1.759	0,94	106	75,18
3	CASES	1.059	0,57	113	80,14
4	HEALTH	769	0,41	111	78,72
5	RESPIRATORY	679	0,36	132	93,62
6	PATIENT	658	0,35	71	50,35
7	VIRUS	594	0,32	91	64,54
8	CASE	552	0,29	101	71,63
9	CORONAVIRUS	550	0,29	87	61,70
10	HOSPITAL	538	0,29	76	53,90
11	CARE	536	0,29	72	51,06
12	DISEASE	529	0,28	100	70,92
13	DAYS	514	0,27	85	60,28
14	INFECTION	466	0,25	100	70,92
15	REPORTED	429	0,23	98	69,50
16	HONG	426	0,23	94	66,67
17	KONG	424	0,23	94	66,67
18	CLINICAL	379	0,20	76	53,90
19	OUTBREAK	365	0,19	84	59,57
20	PROBABLE	365	0,19	73	51,77
21	SEVERE	343	0,18	128	90,78
22	TRANSMISSION	339	0,18	81	57,45
23	CONTACT	324	0,17	76	53,90
24	COV	323	0,17	32	22,70
25	ACUTE	321	0,17	132	93,62
26	SYNDROME	307	0,16	131	92,91
27	SYMPTOMS	298	0,16	76	53,90
28	MARCH	296	0,16	59	41,84
29	DURING	293	0,16	81	57,45
30	CONTROL	281	0,15	88	62,41
31	ASSOCIATED	279	0,15	68	48,23
32	ILLNESS	261	0,14	66	46,81
33	CHINA	247	0,13	81	57,45
34	FEVER	237	0,13	53	37,59
35	VIRAL	233	0,12	50	35,46
36	SUSPECTED	227	0,12	55	39,01
37	TORONTO	225	0,12	53	37,59
38	HUMAN	220	0,12	50	35,46
39	INFECTED	217	0,12	75	53,19

40	TRAVEL	214	0,11	65	46,10
41	USED	208	0,11	56	39,72
42	APRIL	207	0,11	54	38,30
43	PNEUMONIA	203	0,11	63	44,68
44	CHEST	199	0,11	41	29,08
45	IDENTIFIED	198	0,11	70	49,65
46	LABORATORY	196	0,10	60	42,55
47	AMONG	195	0,10	57	40,43
48	PUBLIC	192	0,10	59	41,84
49	HIGH	191	0,10	59	41,84
50	CELLS	190	0,10	31	21,99
51	NUMBER	189	0,10	76	53,90
52	PERSONS	189	0,10	39	27,66
53	DATA	188	0,10	62	43,97
54	SPREAD	188	0,10	80	56,74
55	INCLUDING	184	0,10	87	61,70
56	WORKERS	184	0,10	66	46,81
57	PCR	183	0,10	30	21,28
58	HOWEVER	182	0,10	78	55,32
59	ONSET	180	0,10	50	35,46
60	THREE	180	0,10	62	43,97
61	SAMPLES	179	0,10	36	25,53
62	CDC	172	0,09	40	28,37
63	DAY	172	0,09	52	36,88
64	FIGURE	171	0,09	42	29,79
65	INFECTIOUS	171	0,09	62	43,97
66	TREATMENT	171	0,09	53	37,59
67	RIBAVIRIN	170	0,09	31	21,99
68	USE	166	0,09	57	40,43
69	C	162	0,09	43	30,50
70	BOTH	160	0,09	69	48,94
71	ALTHOUGH	159	0,08	80	56,74
72	MIGHT	159	0,08	59	41,84
73	BECAUSE	158	0,08	70	49,65
74	INDEX	157	0,08	37	26,24
75	BETWEEN	155	0,08	50	35,46
76	STATES	155	0,08	30	21,28
77	UNITED	151	0,08	36	25,53
78	DEFINITION	150	0,08	53	37,59
79	INFORMATION	148	0,08	61	43,26
80	MEDICAL	148	0,08	54	38,30
81	SPECIFIC	148	0,08	50	35,46

82	ISOLATION	147	0,08	61	43,26
83	ANALYSIS	146	0,08	33	23,40
84	AVAILABLE	145	0,08	63	44,68
85	WORLD	145	0,08	83	58,87
86	AREAS	144	0,08	53	37,59
87	KNOWN	144	0,08	62	43,97
88	VENTILATION	144	0,08	36	25,53
89	EXPOSURE	143	0,08	37	26,24
90	SERUM	143	0,08	40	28,37
91	QUARANTINE	142	0,08	34	24,11
92	HOSPITALS	141	0,08	51	36,17
93	WITHIN	141	0,08	55	39,01
94	ICU	140	0,07	15	10,64
95	DEATHS	139	0,07	63	44,68
96	STAFF	139	0,07	39	27,66
97	ADMISSION	138	0,07	29	20,57
98	EARLY	138	0,07	65	46,10
99	ILL	138	0,07	38	26,95
100	MEASURES	136	0,07	52	36,88

Keywordlist for the popular subcorpus

N	*Key word*	*Freq.*	*%*	*RC. Freq.*	*RC. %*	*Keyness*
1	PEOPLE	1.400	0,46	135	0,07	700,91
2	OFFICIALS	835	0,28	46	0,02	535,60
3	YESTERDAY	336	0,11	0		325,23
4	CHINA	1.206	0,40	247	0,13	316,29
5	GOVERNMENT	550	0,18	51	0,03	281,00
6	SAYS	362	0,12	10		277,50
7	TODAY	291	0,10	4		246,96
8	DOCTORS	397	0,13	31	0,02	221,21
9	LAST	465	0,15	49	0,03	220,48
10	BEIJING	570	0,19	93	0,05	192,32
11	CHINESE	468	0,16	61	0,03	191,67
12	FLU	249	0,08	8		185,05
13	CITY	362	0,12	34	0,02	183,59
14	WORLD	684	0,23	145	0,08	171,32
15	SCIENTISTS	285	0,09	23	0,01	156,31
16	DISEASE	1.543	0,51	529	0,28	153,21
17	JUST	292	0,10	26	0,01	152,41
18	HOW	370	0,12	51	0,03	144,97
19	ASIA	283	0,09	27	0,01	142,25
20	VICTIMS	154	0,05	1		138,86
21	DOWN	209	0,07	11		135,99
22	GET	175	0,06	5		133,24
23	FAR	260	0,09	26	0,01	127,19
24	KILLED	167	0,06	5		125,96
25	DOCTOR	192	0,06	11		121,35
26	DEADLY	170	0,06	7		118,98
27	SAY	198	0,07	14		115,29
28	SHANGHAI	153	0,05	5		113,27
29	ORGANISATION	157	0,05	7		107,49
30	OFFICIAL	168	0,06	10		104,73
31	TOLD	199	0,07	18		102,98
32	NEWS	169	0,06	11		101,82
33	COME	176	0,06	13		100,55
34	HONG	1.178	0,39	426	0,23	99,68
35	KONG	1.169	0,39	424	0,23	97,99
36	KNOW	184	0,06	17		94,13
37	GOING	141	0,05	7		93,47
38	HARD	96	0,03	0		92,89
39	ANNOUNCED	124	0,04	4		92,06

40	HIT	137	0,05	7		90,00
41	SPREAD	619	0,21	188	0,10	83,22
42	DIRECTOR	194	0,06	24	0,01	82,56
43	COUNTRY	254	0,08	43	0,02	82,56
44	YEAR	327	0,11	69	0,04	82,34
45	STILL	320	0,11	67	0,04	81,41
46	WARNING	127	0,04	7		81,35
47	WEEK	374	0,12	90	0,05	77,91
48	SPREADING	166	0,06	18		77,27
49	UK	96	0,03	2		77,20
50	OFF	162	0,05	18		74,23
51	CAME	143	0,05	13		73,80
52	BRITAIN	75	0,02	0		72,57
53	CHIEF	114	0,04	7		70,24
54	PARTY	81	0,03	1		69,49
55	SCHOOLS	112	0,04	7		68,55
56	HEALTH	1.752	0,58	769	0,41	68,12
57	WARNED	97	0,03	4		67,86
58	CAPITAL	139	0,05	14		67,70
59	SICK	160	0,05	20	0,01	67,58
60	AROUND	238	0,08	46	0,02	66,89
61	STOHR	69	0,02	0		66,76
62	MONTH	144	0,05	16		65,98
63	STUDENTS	147	0,05	17		65,59
64	PANIC	99	0,03	5		65,27
65	BELIEVED	131	0,04	13		64,36
66	TAKE	168	0,06	24	0,01	63,88
67	AGO	106	0,04	7		63,49
68	FOREIGN	86	0,03	3		62,73
69	GERBERDING	73	0,02	1		61,96
70	THOUGH	104	0,03	7		61,81
71	HERE	153	0,05	21	0,01	60,13
72	GETTING	71	0,02	1		60,08
73	LITTLE	106	0,04	8		59,97
74	THINK	105	0,03	8		59,15
75	GO	141	0,05	18		58,63
76	ALREADY	189	0,06	34	0,02	57,62
77	SCHOOL	103	0,03	8		57,51
78	MASKS	203	0,07	39	0,02	57,45
79	LIU	58	0,02	0		56,12
80	ASKED	101	0,03	8		55,87
81	EPIDEMIC	373	0,12	108	0,06	55,58

82	HEYMANN	100	0,03	8		55,06
83	TRYING	91	0,03	6		54,54
84	CHEN	56	0,02	0		54,18
85	SUFFERING	56	0,02	0		54,18
86	FEARS	94	0,03	7		53,50
87	EXPERTS	173	0,06	31	0,02	52,97
88	PUT	123	0,04	15		52,88
89	THOUSANDS	93	0,03	7		52,68
90	TOLL	112	0,04	12		52,52
91	EAST	84	0,03	5		52,36
92	CALLED	122	0,04	15		52,15
93	SOUTHERN	177	0,06	33	0,02	51,86
94	MYSTERIOUS	68	0,02	2		51,47
95	MR	114	0,04	13		51,34
96	KILLER	53	0,02	0		51,28
97	WOMAN	134	0,04	19	0,01	51,27
98	AGENCY	99	0,03	9		51,09
99	BRITISH	99	0,03	9		51,09
100	BAD	67	0,02	2		50,56

Keywordlist for the medical subcorpus

N	*Key word*	*Freq.*	*%*	*RC. Freq.*	*RC. %*	*Keyness*
1	PATIENTS	1.759	0,94	674	0,22	1.157,89
2	CLINICAL	379	0,20	23		572,84
3	PATIENT	658	0,35	223	0,07	480,97
4	ASSOCIATED	279	0,15	19		412,00
5	CORONAVIRUS	550	0,29	195	0,06	386,86
6	COV	323	0,17	52	0,02	367,76
7	CARE	536	0,29	231	0,08	312,74
8	PERSONS	189	0,10	12		283,05
9	PROBABLE	365	0,19	120	0,04	273,25
10	ONSET	180	0,10	14		258,07
11	HCWS	131	0,07	0		251,14
12	TRANSMISSION	339	0,18	116	0,04	245,60
13	ICU	140	0,07	3		242,17
14	SERUM	143	0,08	5		235,28
15	RESPIRATORY	679	0,36	432	0,14	235,12
16	CASE	552	0,29	312	0,10	230,07
17	PCR	183	0,10	24		225,53
18	INDEX	157	0,08	14		217,64
19	ADMISSION	138	0,07	7		215,24
20	CHEST	199	0,11	37	0,01	212,33
21	CELLS	190	0,10	32	0,01	212,11
22	C	162	0,09	18		210,96
23	SPECIMENS	131	0,07	8		197,66
24	ARDS	101	0,05	0		193,61
25	VENTILATION	144	0,08	15		191,21
26	ANTIBODY	120	0,06	6		187,60
27	VIRAL	233	0,12	70	0,02	186,87
28	FIGURE	171	0,09	31	0,01	184,63
29	LABORATORY	196	0,10	51	0,02	173,52
30	RADIOGRAPH	90	0,05	0		172,52
31	INFECTION	466	0,25	285	0,09	172,12
32	ISOLATES	95	0,05	1		171,96
33	DEFINITION	150	0,08	24		171,16
34	MG	87	0,05	0		166,77
35	EXPOSURE	143	0,08	23		162,81
36	RIBAVIRIN	170	0,09	39	0,01	162,47
37	NOVEL	106	0,06	6		162,21
38	ANALYSIS	146	0,08	25		161,78
39	MECHANICAL	99	0,05	4		159,82

40	RNA	120	0,06	14		153,85
41	SARS	3.001	1,60	3.547	1,18	153,84
42	DATA	188	0,10	55	0,02	153,70
43	TABLE	114	0,06	13		147,23
44	SPECIFIC	148	0,08	32	0,01	146,20
45	DURING	293	0,16	145	0,05	145,81
46	INTERFERON	87	0,05	3		143,37
47	RT	98	0,05	7		143,20
48	POSITIVE	135	0,07	27		138,93
49	DAYS	514	0,27	375	0,12	137,64
50	CONTACTS	134	0,07	28		134,81
51	FEATURES	84	0,04	4		132,35
52	SEQUENCES	95	0,05	9		129,57
53	SAMPLES	179	0,10	65	0,02	123,17
54	CRITERIA	89	0,05	8		123,10
55	COHORT	64	0,03	0		122,68
56	NASOPHARYNGEAL	64	0,03	0		122,68
57	SEQUENCE	119	0,06	24		121,94
58	FINDINGS	134	0,07	34	0,01	120,53
59	MEDIAN	62	0,03	0		118,84
60	USED	208	0,11	92	0,03	117,90
61	NEGATIVE	119	0,06	26		116,87
62	AVAILABLE	145	0,08	44	0,01	115,41
63	INCLUDED	107	0,06	20		113,85
64	CELL	119	0,06	28		112,05
65	H	57	0,03	0		109,26
66	LYMPHOPENIA	57	0,03	0		109,26
67	CONVALESCENT	56	0,03	0		107,34
68	VERO	70	0,04	4		106,93
69	CULTURE	79	0,04	8		105,75
70	ALVEOLAR	55	0,03	0		105,43
71	MAR	55	0,03	0		105,43
72	RADIOGRAPHS	55	0,03	0		105,43
73	METAPNEUMOVIRUS	67	0,04	4		101,52
74	INITIAL	116	0,06	31	0,01	100,91
75	OUTCOME	68	0,04	5		98,72
76	EPIDEMIOLOGIC	51	0,03	0		97,76
77	NUCLEOTIDE	51	0,03	0		97,76
78	USE	166	0,09	71	0,02	97,53
79	DISTRESS	59	0,03	2		97,42
80	G	92	0,05	18		95,72
81	IDENTIFIED	198	0,11	101	0,03	94,82

82	PRIMERS	49	0,03	0		93,92
83	PRESENTATION	57	0,03	2		93,73
84	WWW	57	0,03	2		93,73
85	ML	48	0,03	0		92,01
86	B	91	0,05	20		89,08
87	ELEVATED	51	0,03	1		88,84
88	OBTAINED	71	0,04	9		88,53
89	SERA	46	0,02	0		88,17
90	AGENT	116	0,06	38	0,01	87,03
91	SECONDARY	72	0,04	10		86,88
92	INTRAVENOUS	53	0,03	2		86,34
93	SEROLOGIC	45	0,02	0		86,26
94	MACAQUES	52	0,03	2		84,50
95	INFILTRATES	44	0,02	0		84,34
96	REQUIRED	100	0,05	28		84,30
97	MARCH	296	0,16	210	0,07	83,82
98	INTUBATION	48	0,03	1		83,21
99	INVESTIGATIONS	59	0,03	5		82,84
100	DYSPNEA	43	0,02	0		82,42

Index

English Corpus Linguistics

Edited by Thomas Kohnen and Joybrato Mukherjee

Vol. 1 Roberta Facchinetti / Frank Palmer (eds.): English Modality in Perspective. Genre Analysis and Contrastive Studies. 2004.

Vol. 2 Grace Y. W. Tse: A Corpus-based Study of Proper Names in Present-day English. Aspects of Gradience and Article Usage. 2005.

Vol. 3 Sabine Braun / Kurt Kohn / Joybrato Mukherjee (eds.): Corpus Technology and Language Pedagogy. New Resources, New Tools, New Methods. 2006.

Vol. 4 Clemens W. A. Fritz: From English in Australia to Australian English. 1788–1900. 2007.

Vol. 5 Christiane Brand: Lexical Processes in Scientific Discourse Popularisation. A corpus-linguistic study of the SARS coverage. 2008.